HIGH SPEED CMOS DESIGN STYLES

HIGH SPEED CMOS DESIGN STYLES

by

Kerry Bernstein
Keith M. Carrig
Christopher M. Durham
Patrick R. Hansen
David Hogenmiller
Edward J. Nowak
Norman J. Rohrer

IBM Microelectronics

KLUWER ACADEMIC PUBLISHERS
Boston / Dordrecht / London

Distributors for North, Central and South America:
Kluwer Academic Publishers
101 Philip Drive
Assinippi Park
Norwell, Massachusetts 02061 USA
Telephone (781) 871-6600
Fax (781) 871-6528
E-Mail <kluwer@wkap.com>

Distributors for all other countries:
Kluwer Academic Publishers Group
Distribution Centre
Post Office Box 322
3300 AH Dordrecht, THE NETHERLANDS
Telephone 31 78 6392 392
Fax 31 78 6546 474
E-Mail <services@wkap.nl>

 Electronic Services <http://www.wkap.nl>

Library of Congress Cataloging-in-Publication Data

A C.I.P. Catalogue record for this book is available
from the Library of Congress.

Printed on acid-free paper.

Printed in the United States of America

Preface

In response to the need for a compilation of high-speed CMOS conventions and structures used in the microprocessor design community, the following text is offered. Emphasis is placed on examining the performance variability of these styles contributed by process, design, and use conditions. Comments are based on the authors' direct experience in the crafting and fabrication of CMOS microprocessors used in high performance workstation applications. If the authors have accomplished their goal, the trade-offs in performance, power, area, reliability, and cost made by the selection of design style will have been made readily apparent.

The text is aimed at the graduate level student or engineer mainly interested in circuit design, and is intended to provide some practical reference, or "horse-sense" to mechanisms typically described with a more academic slant. This book is organized so that it can be used as a textbook or as a reference work.

Chapter 1 describes sources of process-driven performance variation in quarter-micron CMOS, and offers rules of thumb for dealing with them. Basic process-related design considerations are developed, and provide a basis for discussion throughout the rest of the book. While these tolerance concepts are not limited to a single fabricator or process, the specifications for a particular operation override the rough rules of thumb provided here.

Chapter 2 surveys non-clocked, static circuit families used to implement combinatorial logic. The operation of each style, as well as its respective strengths and weaknesses are explained. Unique characteristics and sensitivities are addressed in more detail in the "Characteristics" subheading for each family.

Chapter 3 examines clocked and dynamic logic. Features which differentiate each style from its clocked and non-clocked predecessors are identified. Because clocked logic in general and dynamic logic in particular brings with it higher design liabilities, additional effort is made to describe mechanisms known to create headaches.

Chapter 4 explores design-driven performance variability and, along with the process-related variation, discusses the allocation of design margin which accommodates the composite delay tolerance. Important topics such as on-chip device length tolerance, noise, supply rail inconsistency and temperature variation are integrated in this chapter.

Various storage element designs developed for high speed processors are covered in Chapter 5. In Chapters 2 and 3, numerous logic circuit structures were shown to execute logic operations along their paths. Techniques and strategies covered in this chapter use the multiple latch configurations described to first set up data inputs to be evaluated, and then to later capture the calculated outputs in an efficient manner.

An overview of popular chip interface structures and related performance considerations to support high speed chip communications is provided in Chapter 6. Not only must the circuit configurations enable high speed communication; the selected input/output convention must be compatible with the signal levels and design style used throughout the rest of the chip.

Chapter 7 deals with the multiplicity of clocking styles found in the industry. The previous chapters hinted at the variety of clocking styles needed for the operation of specific logic circuits, latches, and I/O devices previous explained. In this chapter, we are now fully introduced to the significance of clocking choice, and its profound influence on performance.

Two important contemporary timing concepts, slack borrowing and time stealing, are introduced in Chapter 8. Significant performance implications are associated with how clock boundaries are defined. These practices are only made possible by the use of newer innovative design styles, building upon the latching and clocking concepts developed in chapters 5, 6, and 7.

Finally, Chapter 9 updates the reader on emerging technology directions which are likely to have a substantial impact on future microprocessor design styles. Certain device design opportunities will enable circuit configurations not presently feasible.

This text is not intended to be a circuit design guide, nor an authoritative reference for the physics of VLSI. Rather, this is a source of good ideas and a compilation of observations, highlighting how different approaches trade off critical parameters in design and process space. Our industry is marked by the contributions of many innovative people; some of their best works are referenced within. The reader is directed to more focused texts and to the cited references for more thorough treatments of specific topics.

These are exciting times to be in this industry. The limits to scaling are now becoming evident, making the need for innovation even more urgent. In a setting where competitors all buy essentially the same fabrication tools and develop quite similar technologies, what differentiates products from one another is the logic configuration selected, and the circuit topologies used to implement it. A design point which merely perpetuates old design styles in a new technology, using existing synthesis and checking tools, is doomed to, at best, matching the competitive disposition it occupied in the prior technology. The selective introduction of superior design styles with enlightened technology usage is imperative for a competitive product.

Acknowledgments

If this text is successful in sharing insight and experience, it is due in no small part to the legion of highly skilled and dedicated employees that the authors have been honored to work among over the years. Although literally hundreds of people have shared their insight with us, a few individuals merit special recognition.

This book originates from a lecture presented at the MIT MTL VLSI Seminar. The topic was cultivated by extended discussions with Dr. Larry Heller. I am indebted to Larry for his insight, generosity, and patience. Ron Black, Director of IBM PowerPC Product Development, recognized the need for a text which "tells it like it is," and encouraged it's development. The job of actually creating such a text would have been impossible without the support and encouragement of Sol Lewin, IBM PowerPC Chief Engineer/Technologist. Thank you, Sol. We have benefitted from the valuable input of Andy Bryant, Rob Busch, Howard Chen, Bill Clark, John Cohn, John Connors, Dennis Cox, Emmanuel Crabbe, John Ellis-Monaghan, Wesley Favors, Frank Ferraiolo, Bill Klaasen, Diane Kramer, Mark Lasher, David Lackey, Henry Levine, Steven Luce, Ed Maciejewski, Tom Maffit, Robert Masleid, Michael Maurice, Daniel Menard, Glen Miles, Steve Mittl, Steven Oakland, Tim O'Gorman, Philip Restle, John Sheets, Yuan Taur, Xiawei Tian, Bill Virun, Sally Yankee, and Jeff Zimmerman. Ione Minot always had a fix for our publishing software snafus, and somehow Bruce Blackman kept making critically needed hardware magically appear.

Finally, I am indebted to my co-authors, who made time in an already heavily-committed schedule, to join me in this endeavor.

Kerry Bernstein
Underhill, Vermont

This work is dedicated to our families, who kept the fire going in the woodstove during this project; and to our professional mentors and peers, who unselfishly share their experience and insight.

Contents

CHAPTER 1

Process Variability

1.1 Introduction

A robust, optimized VLSI design not only matches circuit design style to the architecture and function to be implemented, but also to the CMOS technology in which the design is to be manufactured. It is imperative for the circuit, latching, and clocking scheme to anticipate the fabrication process. In the most desirable scenario, a process is configured during its development to target electrical parameters preferred by the specific design styles intended to be used. However, many a designer is faced with the challenge of providing a product that must be portable among fabricators with differing process details and thus a variety of parameter spaces. In this case more attention must be paid to tolerance of the variation in process conditions. So, before examining popular design practices, we need to look at predominant contributors to performance variation which will affect the various styles in different ways.

We distinguish among three categories of process variation, namely

1. **Fabricator-to-Fabricator Variation** (Inter-Fab Variation): We refer here to *intentional* differences in process that fabless design shops must accommodate with their designs and products. These usually require at least minor modification to designs even when the most-tolerant styles are used. We will explicitly address this type of variation only where inter-fab variation leads to some unique, novel, or unexpected considerations above and beyond those encountered from the second and third categories

2. **Inter-Die Variation**: We refer here to the variation of process (and parametric response) experienced from die-to-die for many samples of many wafers over many lots of fabricator runs. This is the most-commonly considered form of variation in the literature.

3. **Intra-Die Variation**: We refer here to the variation expected *within a single die*. This variation is most-often addressed in the form of device matching but must be considered in a much broader perspective, as outlined below.

Typically, chip fabrication is divided into Front-End-Of-Line (FEOL) and Back-End-Of-Line (BEOL) processing sectors. Simply put, FEOL processing builds the transistors and BEOL processing builds the wires which connect them to one another. Electrical parameter tolerances and fundamental properties of the materials used must be accommodated in the design to assure functionality, wide process window yield, and high reliability. For most purposes, the distribution of electrical values of the two process line sectors are unrelated to each other.

We begin by first examining why process variation is a concern in CMOS logic design. In Section 1.2, the predominant process variation mechanisms associated with FEOL processing are described. While on the topic of FEOL processing, we survey in Section 1.3 process-induced charge-leakage mechanisms, which will be important for selected circuit topologies discussed later in this text. Section 1.4 addresses process variation in the BEOL which can affect performance. Section 1.5 then summarizes the nature of process variation, and the implications for performance.

With that said, let us begin:

1.1.1 Inter-Die Variations: Across-Lot and Across-Wafer Variation

As we will examine in Chapter 4, timing margin in high speed logic designs must be budgeted to accommodate sources of delay variability. Generally speaking, the tolerance of electrical parameters affecting performance can be represented as having lot, wafer, and chip components. Variations across *lots* and across *wafers* are usually not the biggest threat to functionality;   most of the logical circuit paths on a chip react similarly to this shift, and timing-wise, their delay tends to move in the same direction, preserving timing relationships. So why should we worry about tolerance at this level?

1. Functionality needs to be assured at the process extremes.

Chip functional yield is optimized when the design has been verified to function at the ends of the distribution of values of a given electrical parameter. The distribution is commonly due to process variation (Circuit designers love to blame the process!).

Rule of thumb: A standard circuit design practice is to ensure functionality for the +/- 3 sigma *timing delay variation* found by allowing all significant electrical parameters to vary randomly across their own 3-sigma window over many random statistical simulations of a given circuit model.

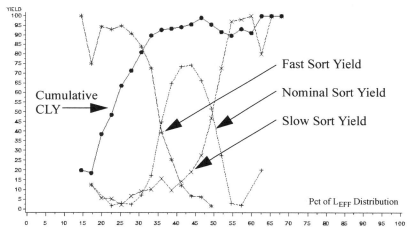

FIGURE 1.1 Sort Percent Yield dependence on process for a non-optimized design

Figure 1.1 shows what happens if these issues are not considered. The plot contains the yield history of an actual product which was designed without regard to functionality across the process window. Mean chip channel length, its value shown as a percentage of the total process window penetration, is plotted against percent yield. The **Cumulative CLY**[1] curve is the total yield of all parts limited by circuit functionality rather than defects or processing errors. Note that the design does not function at *any* speed for a substantial portion of the distribution, yet achieves near 100% functionality at its design point, centered 50% into the L_{EFF} process window. The **Fast, Nominal, and Slow Sort Yield** curves show the percent of the Cumulative CLY which appears in each sort at the given L_{EFF} and

---

1. "Circuit-Limited-Yield"

again shows the design's inability to accept process tolerance. While the percentage of parts fast enough to go into the fast bin rises strongly with short L_{EFF}, the absolute CLY plummets.

2. Response variation can be caused by path composition differences.

   The ratio of intrinsic delay[2] to RC delay[3] of global critical paths[4] can vary widely on larger die. These paths will respond very differently to a particular process parameter. For example, the heavily intrinsic net will not be very sensitive to wire capacitance changes, while a path with an abundance of interconnect loading will change considerably. If two dissimilar paths must maintain a timing relationship, functionality and yield could suffer (Process developers love to blame the design!). Figure 1.2 illustrates the response of three different paths to a hypotheti-

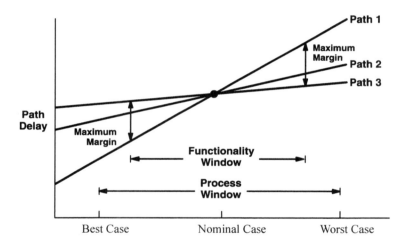

FIGURE 1.2 Effect of varying path composition on functionality

cal process variation. As the paths "disperse" from their centered design point due

---

2. "Intrinsic delay" refers to unloaded CMOS stage delay, the delay of the base logic before the output is capacitively stepped up to drive a large wire and/or fanout gate load capacitance.

3. "RC Delay" refers to the delay incurred driving the interconnect and fanout gate load through the resistance of the interconnect, which becomes large at the global interconnect level.

4. "Critical Paths" refers to the slowest logic paths on chip, which become the cycle-speed-limiting operations. The clock operating frequency of the chip is determined by these paths.

to process variation, timing margin between them is consumed until the design ceases to function. Chapter 8 explores timing in more detail, and Chapter 4 examines design margin allocation.

> **Rule of thumb:** On a large microprocessor, the composition of delay can range from 95% intrinsic delay and 5% path RC delay[5] to 40% intrinsic delay and 60% RC delay.

1.1.2 Intra-Die Variation

Parameters with on-chip tolerances which do not necessarily track chip or wafer mean process are of particular concern. This tolerance element causes paths, even with similar compositions and on the same chip, to vary in their timing relationship. As we will explore in Chapter 4, short paths are most at risk because there are fewer opportunities to "amortize" out the variation. The path's built-in margin must accommodate most of this type of tolerance, unlike the tolerance associated with across-wafer or across-lot, because it moves *against* other paths. Contributors to intra-die variation are considered in "Front-End-Of-Line Variability Considerations" on page 6.

1.1.3 Fail Causes

So, given an awareness of how tolerance threatens the functioning of the design, where do we need to be vigilant?

1. Race conditions

   If two or more independent logic paths, both launched by the same clock, have required timing relationships which are not controlled by subsequent clocks, then the designer must ensure that in each process corner, this relationship is preserved. An example is one signal which must always arrive at a certain logic gate before another. Similarly, when logic signals are used as virtual clock pulses to gate or enable subsequent logic paths, then in the worst combination of allowable conditions, the enabling input must still arrive in time to qualify the secondary path for evaluation. Timing is covered in more detail in Chapter 8.

2. Conventional Timing Margin

---

5. Path RC Delay refers to the time required for a circuit's output signal to drive its target load through the interconnect wiring.

Conventional timing margin refers to the standard delay assessment for a logic path. Delay rule timing attempts to guarantee sufficient time between the launch and capture clock to complete evaluation or preconditioning. Again, see Chapter 4.

3. Clock Skew and Jitter

Inaccuracy in the position of clock timing edges due to variations in clock regeneration behavior can cause the same effect as circuit delay variation: insufficient evaluate or preconditioning time provided to the logic circuitry will cause fails. Chapter 7 describes in more detail the nature of clock signal delivery.

4. Circuit Fails

Certain circuit topologies have very specific electrical parameter requirements. Instances of these styles must be modeled to assure functionality when the given process wanders towards the family's sensitive corner(s). For example, if PFET threshold voltages go too low in DCVS[6] logic, it will become difficult for the logical inputs to switch the output. Chapter 4 examines the sensitivities of specific circuit styles.

Next we examine the specific mechanisms which contribute to delay variability.

1.2 Front-End-Of-Line Variability Considerations

Front-End-of-Line (FEOL) variability affects the response of discrete electrical components (FETs, capacitors, resistors) fabricated in the silicon. Figure 1.3 illustrates the basic FET layout and its dimensional sources of variation. The FET is strongly influenced by variation in its effective channel length (A), poly gate length (B), spacer widths (C), and gate oxide thickness (D). To a lesser extent device performance is affected by device width/edge effect variations (E). These effects are explained below.

1.2.1 Short Channel Effects and ACLV

Most circuit design styles are very sensitive to channel length and variation.

One usually strives to minimize channel length to achieve best performance since both "on" resistance and gate capacitance are largely determined by this parameter.

---

6. See "Differential Cascode Voltage-Switched Logic (DCVSL)" on page 59.

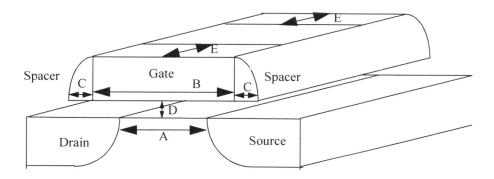

FIGURE 1.3 Basic FET structure, and sources of variation

As the distance from source to drain is reduced, the potential barrier established electrically between the source and drain diminishes in height, and presents lower device threshold voltage, increased dependence of threshold voltage on drain voltage, and increased subthreshold swing. These phenomena, collectively known as *short channel effects* (SCE), exaggerate the electrical impact of systematic channel length variation, more-so than a part with longer average channel lengths [1.1]. Relatively minor changes in channel length on fast (short-channel) hardware can cause substantial threshold voltage changes.

> **Rule of thumb:** In the subthreshold region of operation, drain-to-source current can be expected to increase by 10X for every 80 to 100 mV of gate bias for operation at temperatures between 25°C and 100°C.

Threshold voltage dependence on channel length is plotted in Figure 1.4. Because channel length has a first order influence on performance, the magnitude of its tolerance is critical. Figure 1.5 shows the impact on static chip performance of 3-sigma process variation of channel length along with threshold voltage and gate oxide.

Channel-length variation is created by a combination of photolithography, gate etch, spacer formation (deposition and directional etch), ion implantation and thermal processing. Tolerance in the size of the mask image defining the gate, the angle of the etched polysilicon gate sidewall, thickness of the "spacer" applied on the sidewalls of the gate, and the temperature at which the implants are "activated" all contribute to wider distributions of channel length. Subtle fabrication characteristics such as angle of the ion implant or the concentration profiles of the diffusions and halos cause sub-

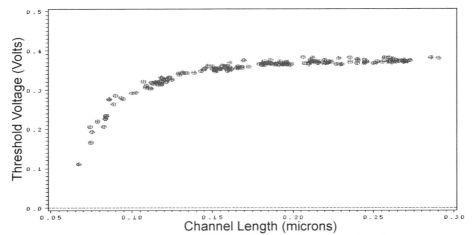

FIGURE 1.4 NFET Threshold Voltage dependence on channel length

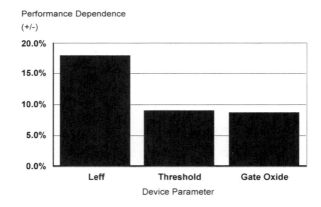

FIGURE 1.5 3-Sigma-Process Performance Dependence

stantial changes in device characteristics. Many articles in the literature describe entire FEOL processes and sources of its variation [1.2] .

There are two major schemes in use for MOSFET construction worth detailing for the purpose of Inter-Fab Variation, as these two methods result in qualitatively different responses of the MOSFETs to process variations.

1. Simple Drain Devices are built by implanting all of the source/drain species after a relatively large (~ 15% of L_{GATE}) spacer has been deployed aside the gate. This device scheme is subject to larger gate/drain overlap variations since the large out-diffusion of a deep source/drain must be compensated by the large spacers and the two process variations are independent and non-cancelling.

2. LDD or Drain-Extension Devices are built by implanting a relatively light (~10^{14}cm^{-2}) source/drain doses immediately after a narrow (~5nm) spacer (either by oxidation of the gate-electrode edge, deposition and etch of a thin film, or both); a relatively large spacer (~20% of L_{GATE}) is then added followed by deep junction implants to allow low-resistance, low-leakage contacts (usually via a self-aligned silicide process). This device scheme frequently will include *halo* (or *pocket*) implants to further reduce short-channel effects.

Channel-length tolerance can be characterized as two separate distributions. *Chip mean channel length*, measured electrically, is the root-mean-square channel length measured lot-to-lot, wafer-to-wafer, and chip-to-chip. This variation describes the time-dependent statistical variation found in a typical manufacturing process, and any associated systematic variation. *Across-Chip Line-width Variation* (ACLV), is the variation about the mean which is found when sampling a large population of gates on a single die. Both will be considered Gaussian distributions. Their relation is shown in Figure 1.6.

> **Rule of thumb:** For a mature VLSI fabrication process, the variation in effective channel length within a single die is roughly 10% of the mean channel length. In less-mature technologies this may easily be doubled.

ACLV is a critical design parameter. It limits the precision with which the designer

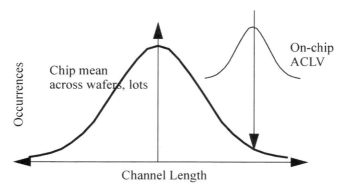

FIGURE 1.6 Gate Line Width variation, on-chip distribution about its mean.

can predict the delay of a circuit path. There are layout precautions which can be taken to avoid the full ACLV tolerance, in cases where good device matching is required. Channel-length variation is specific to the particular process, but in general has been noted to be lower when the device gates:

1. have the same physical dimensions, in both length and width,

2. are oriented in the same direction,

3. are close to one other (within a few hundred microns),

4. have the same nearest-neighbor distances as each other and

5. are placed in areas with similar overall pattern density.

Circuit styles which depend on close NFET-to-NFET or PFET-to-PFET channel length matching, such as cross-coupled latches, should be designed within these guidelines as a minimum.

1.2.2 NFET to PFET Length Tracking

The electrical relationship between NFETs and PFETs has further process variation.

In addition to the channel length variation between like devices described above, there is an additional distribution in the variation between NFET and PFET lengths, or *N-to-P length tracking.* This variation arises from dose, energy, and out-diffusion tolerances of the dopants associated with the different MOSFET types; in some process schemes further differential between NFET and PFET channel length may be caused by the physical morphology of the gate electrode and/or spacers. This variation should be added to ACLV to determine the total NFET-to-PFET length variation. Ratioed circuit design styles are most affected by N-P tracking and ACLV. Relative N-P device strength modifies noise margin and switchpoint in a number of styles.

It is important to note that differences in the very nature of holes vs. electrons in silicon makes 'tracking' of PFETs to NFETs impossible. A first order difference is generated by the combination of near-equal saturation velocities for the two carrier types with low-field mobilities for electrons that are roughly four times that for holes. Figure 1.7 illustrates the resultant dependence of I_{DSAT} on L_{EFF} for PFETs vs NFETs; the PFET I_{DSAT} changes at two-thirds the (absolute) rate of the NFET despite the fact that the PFET has a base value at about half that of the NFET. Thus the ratio of drive strength, I_{DSAT-N}/I_{DSAT-P} varies significantly with channel length. Temperature and voltage may also drive further changes in the drive ratio for similar reasons.

For best PFET-to-NFET tracking specifically,

• The devices should retain as constant a relationship in device length as possible, to reproduce the idiosyncrasies associated with outdiffusion under the gate.

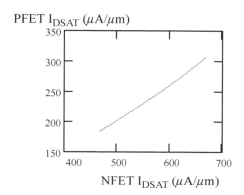
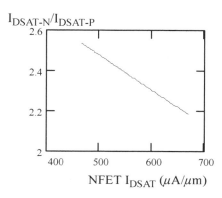

FIGURE 1.7 NFET and PFET I_{DSAT} variation with L_{EFF}: (a) PFET vs. NFET I_{DSAT} as L_{GATE} is varied. Note that while I_{DSAT-P} is roughly half that of I_{DSAT-N}, it increases at roughly two-thirds the rate of I_{DSAT-N}, resulting in (b) I_{DSAT}-ratio decreasing with decreasing L_{EFF}.

- The PFET and NFET for which tracking is a concern should be in close proximity, as many of the mechanisms which generate dose, energy and outdiffusion differences have spatial dependence.

- Relative NFET and PFET pattern/layout density, orientation, and nesting should be similar, simply for ACLV considerations.

  Rule of thumb: NFET-to-PFET Tracking tolerance can be assumed to be about 10% of the mean channel length; actual values will be sensitive to process details.

  Rule of thumb: Assume the ratio of NFET-to-PFET I_{DSAT} can vary by about +/- 10%.

  Rule of thumb: Variations are usually driven by variations in run-to-run for critical tools. As a result, the bulk of variations will be from lot to lot.

1.2.3 Channel Width Effects

Channel width effects cause response differences between wide and narrow devices.

Channel width variation has a second order impact on performance relative to the impact channel length variation has. *Narrow Channel Effects*, however, must be considered for circuit topologies using minimum width devices. These regions assume different threshold voltages than that of the central device channel in a manner that depends on the isolation technique used to electrically separate transistors, Figure 1.8.

A MOSFET exhibiting these edge device effects can be accurately modeled by representing the MOSFET as three devices in parallel (Figure 1.8d).

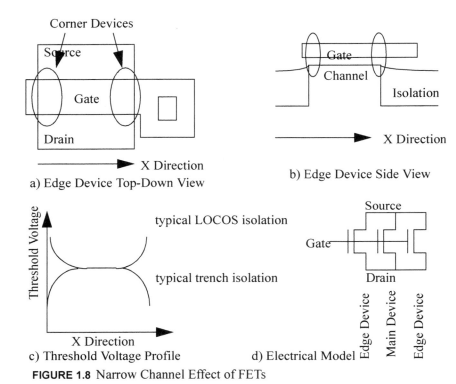

a) Edge Device Top-Down View

b) Edge Device Side View

c) Threshold Voltage Profile

d) Electrical Model

FIGURE 1.8 Narrow Channel Effect of FETs

> **Rule of thumb:** Narrow Channel Effect should be considered in devices narrower than ten times the technology scale length.

In narrow devices, this effect is pronounced. *LOCOS-isolated*[7] minimum width devices typically see *higher* composite threshold voltages *than* wider devices of the same length although in some process integration schemes this sign may actually be

---

7. LOCOS is shorthand for "local oxidation of silicon," which was the predominant means of isolating FETs from one another until the advent of Shallow Trench Isolation (STI). In LOCOS isolation, patterned islands of silicon dioxide are formed between diffusions of adjacent gates. Because of (a) the high "birds-beak" perimeter capacitance and (b) larger required spacing between diffusions, LOCOS is considered inferior to STI.

reversed due to details of placement of well doping with respect to the local-oxidation edge; *Trench-Isolated* minimum-width devices, on the other hand, typically see reduced threshold voltages due to geometrically increased in gate-to-inversion-layer capacitance at the device edges; Many process-specific variations within trench isolation are, however, available to minimize or even reverse the sign of this effect [1.3]. High performance circuit styles use small devices and should anticipate width dependence; when faced with the requirement of functioning in several process technologies, extreme care must be exercised in anticipating even the varying **SIGN** of the response from fabricator to fabricator!

> **Rule of thumb:** In all process technologies, narrow devices should be modeled as three devices in parallel, when ultimate accuracy is required. The edge regions of the narrow FET might be represented by a device from 0.01μm to 0.05μm wide with exalted or depressed threshold voltages, according to the process technology.

1.2.4 Device Threshold Voltage Variation

Device threshold voltage variation limits the accuracy of circuit simulations.

Returning to the first topic of this chapter, threshold voltage becomes lower with shorter channels. In addition, the tolerance on threshold voltage is the worst at shortest channels. *Threshold voltage variation* is caused by local perturbations in dopant implant energy, dose, distribution and gate oxide thickness; since short devices additionally have threshold voltage dependencies on doping profile details at the source and drain edges it is expected that these devices will exhibit the greatest variability. Much more threshold voltage variation is observed at shorter channel lengths, shown graphically in Figure 1.9 using production 0.25μm technology data.

Channel length adjustment is usually the only means available to the designer to control absolute threshold voltage variability at a given effective channel length. From Figure 1.4 and Figure 1.9, channel length, and unfortunately, performance, can be exchanged for threshold voltage precision. As we will explore in the next couple chapters, various styles require differing amounts of switchpoint accuracy. Here's an example of where the capabilities of the available processes must be considered when selecting a style.

1.2.5 Mobile Charge

Mobile charge in a device causes threshold voltage instability.

A known concern with many of the photoresist and developer systems in use is the presence of highly mobile metals in photochemicals, most commonly potassium and sodium. Unless manufacturing practices are vigilant, significant concentrations of

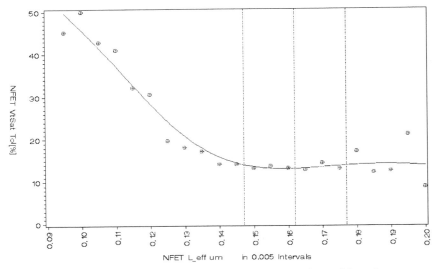

FIGURE 1.9 Threshold voltage tolerance dependence on channel length

these elements can make their way into the gate oxide of the devices [1.4]. Once there, they are easily ionized during use and become mobile positive charges, accumulating in low potential energy regions of the device [1.5]. The natural built-in fields in the device cause these accumulations to occur at junctions and silicon-oxide interfaces, at equilibrium. When bias is applied, however, lower potential energy wells are created which accept the mobile charge and lower the device threshold voltage. Since the device may be used in any number of bias conditions, the low potential energy region of the device varies, which in turn causes the threshold voltage to assume a range of values.

Diagnosing ionic contamination can be quite confusing, since the device, after being allowed sufficient relaxation time, will display normal response. While CMOS process developers strive to make the magnitude of mobile-charge-induced threshold voltage variation insignificant, any known threshold-voltage tolerance regularly caused by ionics must be added to the threshold voltage tolerance specification when modeling performance of a design at process corners of concern.

Rule of thumb: Mobile-charge-induced threshold voltage variability is usually held to less than +/-10% of the threshold voltage of the minimum channel-length device in that technology.

For any given design topology, there is little that can be done to reduce the threshold voltage variation caused by the presence of free charge in the device body. Practice of the matching recommendations made earlier in this chapter can help devices stay matched in the presence of contamination, but unequal stress voltages in use may still result in large differential drifts with time. Eliminating the mobile charge in the device is essential for a stable, robust process.

1.2.6 Hot Carriers

Hot Carrier effects gradually degrade device performance over its lifetime.

Channel carriers (electrons in NFET devices, holes in PFET devices) have the ability to induce damage the host FET structure when excited by high electric fields. The damage accumulates over the life of the product, raising the threshold voltage and reducing drain current in the NFET, and decreasing threshold voltage and increasing drain current in the PFET. This degradation affects performance, and can eventually cause the part to fail when the built-in timing margin of that path is consumed.

There are three modes of hot-carrier-induced degradation: conducting hot-carrier, nonconducting hot-carrier, and substrate hot-carrier.

Conducting Hot Carrier Degradation occurs as the device is turned on. With increasing gate drive, the inversion region forms across the device channel. For gate drives below a critical voltage, the inversion layer is "pinched-off" at some potential, V_P, less than V_{DRAIN}. At pinch-off, the inversion region does not reach the drain. The electric field across the short uninverted remainder of the channel is then extremely high and effectively 'heats' the electrons to a high effective temperature, resulting in a distribution of kinetic energies with a high-energy tail. This scenario is illustrated in Figure 1.10.

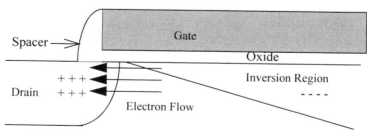

FIGURE 1.10 NFET at pinch-off, and resulting electric fields

While most electrons are collected by the drain (in the case of NFETs), a portion of the carrier energy distribution which is in excess of the gate insulator potential barrier height will be injected into the insulator. In the case of NFETs, most of these electrons will be conducted to the gate electrode. However, some damage is left at the silicon interface in the form of surface states, which results in decreased mobility and degraded subthreshold swing. Figure 1.11 schematically illustrates this process.

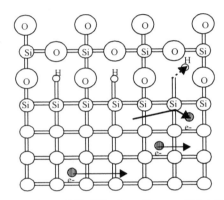

FIGURE 1.11 Representation of Si-SiO$_2$ Interface and Hot Carrier Event

A second mechanism, *charge trapping* of the energetic carrier in the insulator oxide, causes the charge to become resident, resulting in less channel inversion for the remaining life of the product (i.e. V_T increases). Because the damage occurs asymmetrically at the pinched-off (drain) end of the channel, devices such as pass gates that are operated with bidirectional current (see "Complementary Pass Gate Logic (CPL)" on page 73) can see substantial differences over time in one direction compared to the other.

> **Rule of thumb:** For NFETs conducting hot carrier degradation is worst when $V_{GATE}=V_{DD}/2$ and $V_{DS}=V_{DD}$. Slow slew rates through this operating point, or high transistor switch factors cause larger accumulated degradation.

> **Rule of thumb:** Of all the hot carrier degradation mechanisms, Conducting Hot Carrier degradation has historically produced the most severe degradation, and can increase the cycle time of the chip by approximately 1-5% by the chip's end-of-life.

Non-conducting hot carrier degradation is ongoing device degradation which occurs when V_{DD} is across the device but the gate voltage is under V_T. Caused by the same mechanisms as conducting hot carrier degradation, a portion of the inversion carriers present in subthreshold or punchthrough current has sufficient energy to produce damage in the FET gate insulator oxide. The effect is most often seen in shorter channel devices with low threshold voltages and high lateral electric fields. Because the lower threshold voltages motivated by scaling will drive future CMOS technologies to accept higher leakage currents, it may be expected than non-conducting hot carrier degradation will become a greater carrier-caused degradation mechanism.

In PFETs the mechanism is fundamentally the same as that of NFETs, however the reversal of polarities results in some asymmetries. First, the barrier height for holes to enter silicon dioxide is about 1eV higher than that for electrons and so, generally hot electron injection dominates PFET hot-carrier degradation (vs. the naive expectation of hot-hole injection). Second, since the field in the oxide opposes conduction of these injected electrons to the gate electrode in this case, trapping of the electrons in the oxide typically dominates the degradation. Of course the V_T still becomes more positive with PFET hot-electron aging and thus the PFET is more-easily turned on, in contrast to the NFET. More-aggressive fields in advanced technologies can also lead to appreciable hole injection as well, considerably complicating the analysis of hot-carrier degradation in PFETs.

Substrate hot carrier degradation does not involve inversion carrier flow nor high lateral fields. Damage is caused to the gate oxide by carriers energetically injected from the substrate or channel depletion region by high vertical fields. A fraction of the charge present in the bulk or generated in the channel will have sufficient energy to overcome the gate oxide potential barrier and enter the insulator. Most of this charge makes its way to the gate and shows up as gate current. The remainder of the charge fills vacant surface states and becomes resident charge, altering the device threshold voltage. As in the case of channel hot electrons, some surface states are also generated, resulting in degraded channel characteristics. Since thermal electron-hole-pair generation increases charge injection into the insulator, substrate hot carrier degradation becomes significantly more severe at higher temperatures. Because source and drain diffusions also attract away the substrate carriers, substrate hot carrier degradation decreases with shorter channel lengths. Depending on choice of gate oxide, substrate hot carrier degradation is typically the smallest contributor of the three hot carrier degradation mechanisms.

> **Rule of thumb:** For process technologies designed near the scaling limit for their power supply, when V_{DD} is increased (to achieve higher performance) by more than about 20% above the nominal operating voltage for the technology, the end-of-life performance can actually decrease rather than increase, as the performance improvement from higher supply voltage is exceeded by the wearout caused by that voltage.

Hot carrier degradation is a reliability consideration. Because of the complexity required to determine how much each circuit will slow down, it is not practical to re-time a design with each circuit's specific degraded performance. Approximations should be used to anticipate how much performance guardband is required on the part during final test to assure functionality at end-of-life[1.6]. Hot carrier guardband, along with tester-to-tester differences and tester-to-actual-system mismatch must all be combined to determine the total performance margin that must be added to guarantee a given performance when testing the finished chip or module. Degradation is, therefore, very expensive and directly reduces design performance.

> **Rule of thumb:** Hot carrier guardband can be up to 50% of the full additional margin required at final product test needed to reliably deliver a given performance.

Hot carrier degradation can be minimized by prudent design, conservative application conditions, and "safe" process corners. Design styles which minimize the capacitance to be switched, and which appropriately power the driving buffers to their capacitive loads will have less wearout. Similarly, less severe application conditions (lower V_{DD}) and process-centering[8] for lower chip-performance tester sorts (longer channels and higher V_Ts) will also produce longer comparative lifetimes for a given guardband.

> **Rule of thumb:** For calculating general hot carrier degradation across logic, array, clocking, and latch circuits of a whole chip, historically a circuit switch factor of 0.25 has been assumed.

Process solutions to reduce hot carrier degradation have also been offered. Drain engineering, gate-oxided nitridation by various means and more recently, treatments with deuterium [1.7] have been used to reduce channel-hot-electron wearout.

It is interesting to note that for a given device structure, a thicker gate oxide will contain more charge trapping sites, and so has the potential for more conducting degradation than a thinner gate oxide. On the other hand, the thinner gate oxide reduces the device threshold voltage and allows more subthreshold current, making the FET more prone to non-conducting wearout. A thinner gate oxide also has a higher drive current, making available more hot electrons.

Hot-carrier-limited lifetime can be traded for performance. A part which requires long service life with high reliability (for example, 100,000 Power-On-Hours with less than 25 ppm failing, for a microprocessor used in air-traffic-control) can not afford to

---

8. Although technologies are specified for operation at a particular point with a fixed standard deviation, often learning, over a period of time, allows the width of the distribution to be reduced. The manufacturing engineer may choose to "push" the process to the faster end by precisely that amount of learning, so that the same number of parts are still out at the fast end of the distribution. In other cases, even without learning, the manufacturing engineer may still choose to push the process faster to enrich the population of high-speed-sort parts, albeit at the expense of causing some parts at the tail of the distribution to be too fast and either non-functional or unreliable.

accumulate damage, and so must be operated conservatively, or be fabricated in a safer technology. Conversely, a consumer product such as a home computer may require only a 10,000 Power-On-Hour lifetime and so can be designed and operated somewhat more aggressively.

1.2.7 Drain Resistance Modulation

Trapped charge in diffusions adjacent to devices can reduce device current.

A mechanism similar to hot carrier degradation occurs out at the drain edge of the device. During fabrication, a moderately doped *extension*, or *lightly doped drain* (*LDD*) is implanted in the source and drain ("S/D") regions, and is driven under the existing gate structure. The effective channel length is then shorter than the actual physical gate.   Adjacent to each polysilicon gate edge, a *spacer*, composed of an insulator (usually nitride and/or oxide) is then formed, which is used to displace suc-ceeding S/D implants from the gate and prevent silicidation shorts between diffusions and gates (see Figure 1.10). This spacer material (typically CVD silicon dioxide or silicon nitride) tends to have substantially higher trap densities than that found under the gate. If the polysilicon gate overlap of the extension or LDD is insufficient, then charge can become permanently trapped in the spacer oxide beyond the gate as the device is initially stressed, causing depletion of the LDD dopant. Since the gate has little electrical influence over the LDD portion lying beyond the gate, injected charge permanently decreases Ids by increasing the series resistance of the device in the LDD pathway [1.8]. In a well overlapped device, any spacer charge trapping is sus-pended over heavily implanted S/D regions, causing few electrical problems.

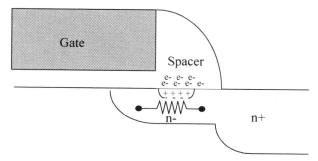

FIGURE 1.12 Series Resistance Modulation in FET Drain Region

A minimum required gate overlap of the LDD or extension must be guaranteed to avoid series resistance modulation[9]. To accomplish this, in the nominal setting, there is substantial gate overlap capacitance to drain and source. This hurts performance, as it is adding Miller Capacitance. This capacitance unintentionally couples input to output, "feeding through" noise and false switching in static circuits, detrimentally "bootstrapping" precharged nodes in dynamic circuits, and generally adding load to both inputs and outputs. These issues will be discussed in detail in Chapter 3.

> **Rule of thumb:** Over a whole chip, the front-end component of fanout loading can be estimated to 66% gate capacitance and 33% S/D capacitance. Of the gate capacitance, approximately 75% is gate-source and gate-drain overlap capacitance, and 25% is gate-substrate capacitance when the device is off; when the device is on, the gate component is virtually 100% inversion capacitance.

While device designers attempt to entirely avoid drain resistance modulation, manufacturing variation can occasionally lead to the presence of this mechanism. The designer should also be vigilant for situations where the performance penalty associated with increased overlap capacitance exceeds that of possible increased series resistance on some percentage of FETs on chip.

1.2.8 Negative Bias Temperature Instability (NBTI)

An aging mechanism similar to hot carriers should be exposed prior to shipment

While not as well understood as hot-carrier degradation, NBTI is an observed aging mechanism. The extended presence of negative voltage on the gate (with respect to the body of the MOSFET) is suspected of enabling an electrochemical reaction which slowly creates additional trap sites at the silicon-oxide boundary[1.9]. Dangling atomic bonds at the silicon-silicon dioxide interface are typically filled during fabrication by hydrogen, as shown in Figure 1.11. Events which uncover these sites or create new sites generate additional locations of fixed positive charge; these trap sites increase the overdrive voltage needed on the gate to achieve equivalent drain-to-source current. NBTI-related trapped charge is observed to have a saturation value which is linearly dependent on the electric field seen by the device; stressing parts with high voltage before shipment achieves the saturated degradation and allows testing with full NBTI degradation. NBTI threshold voltage shifts can be substantial, and so should anticipated. NBTI is highly sensitive to the materials used in and details of the fabrication process. P-doped gate electrodes are more prone to NBTI than n-

---

9. Lightly Doped Drains (LDDs) and extensions are regions formed at the device source and drain by implantation and diffusion under the gate channels edges. The positioning of these structures allow the device to assume a shorter channel length than what lithography alone would support.

doped, for example. This is an area which requires awareness of *inter-fabricator variation*.

> **Rule of thumb:** Fold expected NBTI threshold voltage shift in with hot electron wearout in budgeting total margin required. Because NBTI results from negative gate-to-body bias, it is primarily a consideration for PFETs.

1.2.9 Body Effect

The threshold voltage of a device is altered by its source-to-substrate bias voltage.

Body Effect refers to the effect that elevated source voltage has on the threshold voltage of the device. Figure 1.13 shows an example of a topology which can suffer from body effect. Device T2 is coupled to ground through device T1. Both have identical widths and lengths. Assume current is passing through both devices to ground, as a capacitive load above device T2 is discharged. Due to the voltage drop across the "on" resistance of device T1, the potential of device T2's source is elevated above ground while current is passed. Device T2, in absolute terms, is seeing a substrate

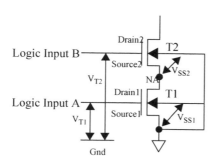

FIGURE 1.13 Body Effect in sequential devices

voltage which is negative with respect to its source voltage (Voltage V_{SS2} is non-zero), which increases the width of the charge depletion region surrounding Source2. The potential across the junction formed between device T2's channel and the substrate increases. This has the effect of making T2's channel region harder to invert to its "on" state. Hence, higher source-to-substrate bias V_{SS2} increases threshold voltage V_{T2}, and the circuit loses performance. Because $V_{SS1}=0$, threshold voltage V_{T1} is not degraded, and device T1 can receive full overdrive.

Unlike other mechanisms is this chapter, body effect is not related to the tolerance on a particular parameter although it is a function of L_{EFF}, with short devices suffering substantially less body effect than long. Nonetheless, this mechanism affects the performance of an affected logic stage, and so must be considered.

> **Rule of thumb:** Assume the threshold voltage of a transistor whose source is coupled to ground through other transistors has ~1.5X the magnitude of threshold voltage it would have had with its source tied directly to ground.

1.2.10 Other Process Parameters

Effective gate oxide thickness is dependent on gate doping concentration.

Gate oxide thickness has a first-order impact on FET performance. It directly affects transconductance, device threshold voltage, device current, and hence chip performance[1.10]. Gate insulators are thermally grown, and modern fabrication processing exhibits a fairly tight tolerance on its physical thickness.

The effective gate oxide is the electrically-equivalent ideal insulator thickness. As the gate of an NFET is raised to V_{DD}, a depletion region is intentionally formed in the channel region of the well. On an NFET device, a second, unintended depletion region forms in the gate electrode immediately above the oxide interface [1.11]. This depletion region acts to make the gate look farther away, or for the device to have a thicker effective gate oxide. The penetration of this depletion region up into the poly-silicon gate is a function of the active gate dopant concentration. Variation in the concentration of the n+ dopant for a typical manufacturing process, is driven predominantly by a shift in the temperature of the anneal which activates the dopant, and translates into variation of effective gate oxide thickness. Single work-function gate technologies are largely immune to these effects due to the ability to keep the gate electrode heavily doped n+.

The result is that the tolerance of the effective gate oxide has a component associated with physical oxide variation, and another coming from implant and anneal variations. In addition, there will be a certain mistracking due to the different dopant species used in NFETs and PFETs. Effective gate oxide thickness tracking for NFET-to-NFET, PFET-to-PFET, and NFET-to-PFET is typically good both locally and across chip. So while gate oxide tolerance must be considered when designing for functionality across the entire process window, it is a second-order contributor to on-chip delay variability.

> **Rule of thumb:** Assume +/- 5% variation in gate oxide thickness across wafer lots over an extended period of time. Very little gate oxide thickness variation is observed across a given wafer.

Silicided gate and diffusion resistance can cause significant delay variation.

The resistivity of polysilicon gates and S/D diffusions may be reduced by the use of a self-aligned silicide (salicide). In the salicide process a metal, such as titanium or cobalt, is first deposited on the surface of the wafer. The wafer is then heated to a sufficiently high temperature for the metal to react with the silicon in polysilicon gate and S/D diffusion to form a silicide. During this process the metal does not react with the oxide or silicon nitride that may be present in the device isolation regions or gate sidewall spacer. The unreacted metal may then be selectively etched, leaving the low resistivity silicide only on the polysilicon gates and S/D diffusions.

Titanium silicide exists in two crystallographic phases, which differ is resistivity by approximately a factor of 4 [1.12]. The high resistance phase may form first, requiring transformation to the lower resistivity phase by a high temperature anneal. Transformation to the lower resistivity phase is nucleation limited, and can be incomplete on narrow structures. This may result in a resistance distribution with two populations, one centered about the high resistance phase and the other centered on the low resistance phase.

The impact of this resistance variation depends upon the particular circuit layout, but it may result in a considerable variation in the RC delay of polysilicon gate wiring. Wide devices contacted only on one end of the S/D diffusion may experience overdrive loss due to resistive voltage drop along the source (Figure 1.14a). Likewise, contacting only on one end of a long gate will result in increased gate charging delay.

(Figure 1.14b). As will be discussed in "Defect Leakage" on page 32, silicide can also create substantial leakage paths.

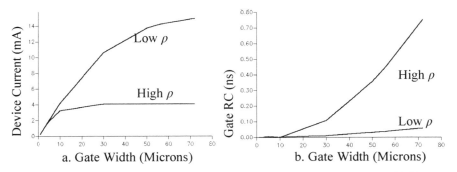

FIGURE 1.14 Overdrive loss and Gate RC versus Gate Width, due to Silicide tolerance on wide FETs.

> **Rule of thumb:** In a process susceptible to non-uniform silicided sheet resistance distribution caused by materials like titanium silicide, local sheet resistance may vary by an order of magnitude. The sheet resistance variation is randomly distributed, varying even among adjacent devices.

The performance impact of this variation mechanism is minimized by defining chip images[10] which do not require use of extended polysilicon gate or diffusion lengths in the conduction path. In this scenario, the dependence on silicide resistance is limited to the short path between the device source or drain and the contact to the interconnect in or out.

Silicides are prone to physical deformation at high process temperatures. The silicide grains have a tendency to separate, which causes the low resistance path through the silicide to become discontinuous. This grain separation is called *agglomeration*. The resulting increase in resistance may be exacerbated by the formation of back-to-back-Schottky diodes at the silicide/doped Si interfaces. There is little that can be done from a design standpoint to avoid this defect, although the occurrence of agglomera-

---

10. "Chip Image" refers to the on-chip power, ground, and clock distribution wiring scheme which allows standardized design of circuit "macros." These macros can then be dropped into the structure with *autoplacement* software where ever needed, and make contact with all necessary support connections.

tion can be controlled by limiting high process temperatures after formation of the silicide.

> **Rule of thumb:** Agglomeration defects in silicided source or drain diffusions or polysilicon gates can create local resistance values as high as several $k\Omega$

1.3 Charge Loss Mechanisms

Alternative high speed circuit styles have sensitivities to technology characteristics which were not previously of major concern. These sensitivities place additional demands on process control and specification. One particular feature, leakage, is a major new concern for a number of reasons:

- A predominant high speed style, dynamic logic, has a particular vulnerability to charge loss. Precharge retention is directly related to noise immunity and functionality.

- In all technologies, increased leakage is associated with reduced technology reliability, caused by nonconducting hot carrier damage, latch-up, etc.

- Some product markets, such as portable electronics, can simply not tolerate the power consumption associated with standby leakage currents, regardless of the source.

- Products are often stressed prior to shipment at elevated temperature and voltage. The high power associated excessive leakage prevents the execution of these reliability tests. In addition, screening product for defects by monitoring standby current[11] can be no longer effective.

Defect mechanisms which have fixed magnitudes of leakage become bigger problems with scaling. As scaling reduces node capacitance, charge retention becomes more urgent.

Charge loss comes from circuit design as well as technology, and have a wide range of relative magnitudes. CMOS-technology-related charge loss mechanisms are shown in Figure 1.15, and are explored below; charge loss associated with design sources are identified on page 96

---

11. Standby current, or I_{DD}(quiescent), is the DC power consumption of the chip when the clocks are not running, and the part is stopped at a particular machine state.

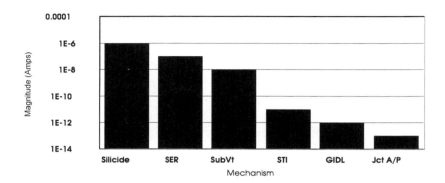

FIGURE 1.15 Relative Magnitudes of Process-related Leakage currents

1.3.1 Subthreshold Leakage Currents

"A nanoAmp here, a nanoAmp there, pretty soon that's a lot of current."

Simple device models depicting the MOSFET in its linear region suggest I_{DS} goes to zero for gate biases under the threshold voltage. This is, in fact, not true in actual hardware. In the subthreshold region of device operation, the drain-to-source current has an exponential dependence on gate voltage. A rigorous explanation of the physics of device leakage current is beyond the scope of this book; a number of excellent treatments can be found in the literature [1.13]. Subthreshold leakage can be described simply as current dominated by the diffusion of minority carriers from the source across the weakly inverted channel to drain, exponentially enhanced by any lowering of the potential barrier between drain and source. The channel barrier is affected by transverse electric field, temperature, and dopant concentrations. Above a relatively low V_{DS} (approximately $2kT/q$) with 0 V gate bias, the effect of the lateral drain-to-source electric field on current saturates, and device subthreshold leakage becomes relatively independent of V_{DS} for long-channel devices. For short-channel devices, DIBL results in and exponentially decreasing barrier near the source, and thus increase subthreshold current.

> **Rule of thumb:** The subthreshold current in a typical submicron process can be approximated to increase one order of magnitude for every 85-90 mV of gate voltage increase towards V_T, at 25°C. V_T will drop roughly 7-10mV for every 10°C increase in temperature, except for N+-gated PFETs where this the V_T will decrease by 12-15mV for every 10°C.

This is a good place to briefly discuss the electrical response of MOSFETs to changes in operating temperature. In Figure 1.16 a semi-log plot of drain current vs. gate voltage is shown for a typical NFET at 25 °C and at 100 °C. Three features are evident. First, V_T decreases with increasing temperature. Gate-electrode composition significantly affects this rate and is a source in inter-fabricator variation when moving between dual-doped and single-doped gate technologies. Second, to make things even worse for leakage, the subthreshold swing increases; the aggregate effect is to significantly increase the subthreshold current. The third effect of increased temperature is that of decreased mobility. Although the mobilities of holes and electrons decrease at similar rates, the impact on drain current is greater for the PFET.

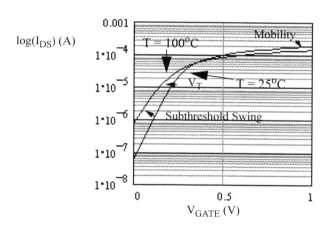

FIGURE 1.16 Temperature response of NFET I_D-V_{GATE} characteristic.

Rule of thumb: V_T will decrease ~1mV/°C with temperature for dual-doped polysilicon technologies. PFETs with n-doped gates will exhibit 50-100% greater temperature dependence.

1.3.2 Junction Leakage

Reverse diode current from the drain to well or substrate can diminish precharge.

Junction leakage refers to reverse current through the diode formed by the junction of the drain diffusion implant and the substrate or well [1.14] . The resulting diode is frequently reversed biased and rarely forward-biased during normal operation. Any free electrons or holes found in the depletion region of, for example, an "n+"[12] to p-substrate junction in reverse bias, will rapidly drift to the n+ diffusion or to the p neutral region, respectively, resulting in a current from diffusion to well. For a p+ diffusions in N-well[13] the same considerations, of course, apply.

This leakage is usually categorized by the source of the carriers, into either *generation current*, or *diffusion current*. Generation current is defined as that due to electron-hole pairs created in the junction depletion zone. These currents are dominant in high-speed CMOS process technologies and are strongly field (and therefore voltage) dependent. Diffusion currents are defined as those contributed by carriers that 'wander' into the depletion zone from the substrate or well. Such currents have negligible voltage dependence but very strong temperature dependence, varying to first order as $e^{-E_{GAP}/(kT)}$, where E_{GAP} is the band-gap energy of silicon. Under ordinary conditions this component of leakage is negligible in logic technologies.

---

12. The notation "p+", "p", "p-", "n+", "n", and "n-", while subjective, is still quite effective in describing relative levels of doping silicon. A common manufacturing approach starts with wafers whose epitaxial layer is *lightly* doped with acceptor ions ("p-"). Since subsequent implantations forming junctions must not only *compensate* the original implant but *reverse* its polarity, a higher doping level must be used ("n"). The notation of "-" indicating a light implant going to "+" referring to a heavy implant is an indicator of relative doping concentration.

13. In a p- epi wafer, p+ diffusions will reside in an NWell biased to V_{DD}. In an n- epi wafer, the p+ diffusions are implanted directly into the substrate which is biased to V_{DD}.

The location of these junctions in a PFET structure are shown in Figure 1.17

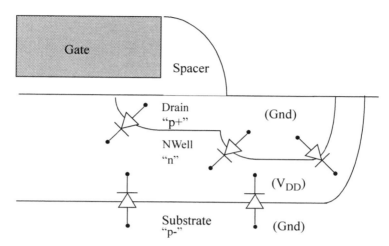

FIGURE 1.17 Built-in CMOS Diodes operated with reverse bias

Rule of thumb: Junction leakage for a typical source/drain diffusion region ranges from tens to thousand of fA.

1.3.3 Field-induced Leakage Mechanisms

Applied electric fields induce leakage paths to power supply and ground.

Gate-Induced Drain Leakage (GIDL) is the result of the strong electric field present at the edge of the drain diffusion under the FET gate, when the gate is in the off state, as shown in Figure 1.18. With the drain at V_{DD}, and the gate and source at ground for an NFET, electric field lines terminate on the gate as well as on the source. This strong field gives rise to leakage through the drain junction perimeter under the gate [1.15]. The relative magnitude of this leakage is usually low, and variation is usually driven by T_{OX} and drain outdiffusion variations. This current is usually due to band-to-band tunneling in the silicon drain region and thus exhibits a small positive temperature coefficient. The process used should be assessed to for the significance of this charge loss path.

Rule of thumb: GIDL for a nominal FET accounts for the loss of no more than a few pA per micron of device width.

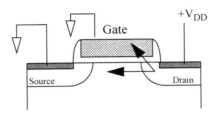

FIGURE 1.18 Gate-induced leakage effects in FETs

1.3.4 Alpha Particle and Cosmic Ray Interactions

Radiation events on chip comprise yet another charge loss mechanisms.

Alpha particles and cosmic rays entering the chip sporadically produce the same result as process-produced leakage paths, depleting existing precharge on dynamic nodes. Alpha particles have extremely short ranges in solids (~ few mm) and as a result must originate in the materials making up the die, its packaging, or associated attach materials. The most common source of alpha particles is from nuclei naturally associated with lead in lead-ball attach technology. Quite amazingly, the other major cause of radiation-induced leakages (for earth-bound VLSI hardware) is due to cosmic rays!

Incident alpha particles strip electrons from the surrounding silicon atoms as they enter the chip. The electron-hole pairs created along the ionizing path are quickly swept up in the prevailing electric fields, as shown in Figure 1.19. For an NFET, the hole goes to ground via the substrate connection, and the electron is attracted to the positively-charged drain, where it crosses the junction and recombines, reducing the existing precharge.

In products using lead alloy solder balls to attach the chip to module, alpha particle flux can be relatively high, but only a fraction of the particles carry the energy necessary to disrupt logic states or memory contents. The particle not only must hit a sensitive node but must do so during the vulnerable "evaluate" portion of its cycle when the node's state is not being actively held. In addition, to cause an error, the data to be sensed must be in the state opposing the "hit" state. One approach to mitigating alpha-induced failures is changing the source of lead used in the solder. It has been observed that lead mined from specific locations has substantially lower alpha emission rates [1.16]. Use of these sources, and avoiding the placement of the solder connections in proximity to vulnerable structures improves the alpha particle contribution to soft

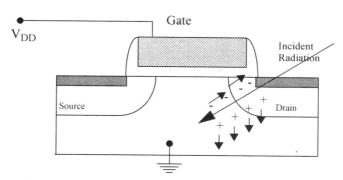

FIGURE 1.19 Radiation-induced electron-hole-pair generation in an NFET

error rate (SER). Lead must be within 30 μm of the silicon node to cause problems. Chips which do not use lead connections generally have alpha particle-induced SER which is lower than that due to cosmic rays.

Actual (primary) cosmic rays are typically protons of extremely high energy but even these high-energy particles rarely can reach the surface of the earth. Instead, mostly third to seventh cascade-generation products [1.17] of collisions with nitrogen and oxygen nuclei in the atmosphere reach our tortured die. While on average one cosmic-induced muon passes through each of our bodies each second, these particles are far too-weakly ionizing to cause much leakage in silicon; rather, neutrons, also generated by nuclear collisions, while much rarer in frequency, if captured by one of the nuclei within our chip, will produce a very large transient leakage when the nucleus subsequently breaks up. Shielding a die from cosmic rays is clearly impractical[14].

To calculate soft error rates, the flux received and the probability of an event must be determined. The flux received is proportional to the vulnerable area and the flux rate of the surroundings. The probability of occurrence is strongly dependent on a) the charge delivery required to upset a logic circuit or memory cell ("Q_{crit}")[15], and b) the

---

14. Although it is worth noting that the Earth's magnetic field significantly alters the cosmic ray flux on Earth; Locations nearer the magnetic poles receive noticeable larger doses of cosmic rays compared to those near the geomagnetic equator.

15. Q_{CRIT} is proportional to the product of V_{DD} and the precharge node capacitance,

depth of the epitaxial layer of the substrate. An expected fail rate may then be determined [1.18].

> **Rule of thumb:** Disturb event rates for SER calculations may be assumed to be:
>
> | | |
> |---|---|
> | 0.005 events/cm^{2}Khr | Cosmic ambient incidence |
> | 0.005 events/cm^{2}Khr | Background alpha ambient incidence |
> | 5 events/cm^{2}Khr | Alpha Incidence from Pb chip connector to package |
> | 10-20 fCoulombs | Charge distribution per alpha event |
> | 100-200 fCoulombs | Charge distribution per cosmic event |
> | 25 fCoulombs | Typical logic Q_{crit} |

Technology scaling reduces the area of the vulnerable nodes by the square of the scale factor which decreases the probability of a radiation hit in proportion. This smaller note, however, will have lower Qcrit due to decreased size (and often voltage). Thus the energy spectrum of cosmic rays must be factored in to understand the net change in cosmic SER. At today's (.25 μm) level of integration, it can be seen that the SER per node is expected to slightly decrease with scaling. The density of nodes goes up by the square of the scale factor, however, and so the SER on a fixed die size (populated with comparably susceptible circuits) increases somewhat more slowly than the scale factor squared. Thus net result is that soft errors are becoming bigger problems in VLSI circuits because while the SER per circuit is roughly constant with scaling, the number of circuits on a die typically increases quadratically.

1.3.5 Defect Leakage

I swear I precharged this node.

Defects in film depositions, annealing, or patterning occur in critical chip process steps due to the presence of foreign material, as well as variations in required process temperatures, times, exposures, or electric/magnetic fields. *Shorts* and *Opens*, with varying severity, are the results of these defects. Shorts, or short-circuit defects, introduce undesired conduction paths between electrical nodes. Opens, or open-circuit defects, conversely, introduce undesired additional resistance in an existing conduction path. Defects can also be characterized as being systematic or random. In "Other Process Parameters" on page 22, we explored polysilicon sheet resistance, which is an example of a systematic "soft" open. Despite perfectly processed hardware, some portion of finished product contains the undesirable high resistance characteristic. Unless gross in occurrence, these defects can pass through production quality screens and still have the potential to cause fails or contribute to wider delay variation. Below are common examples of the many possible random and systematic shorts which rob dynamic node precharge.

Silicides introduce additional junction leakage modes.

Silicides formed on diffusion surfaces have the ability to compromise junction integrity. The formation of the silicide consumes some of the implanted silicon leading to effectively shallower junctions, from a leakage point of view. Some metals (such as titanium) form silicides through diffusion of silicon into the metal. Others (e.g. cobalt) diffuse into the silicon and react to form a metal silicide [1.19] as shown in the source diffusion of the device shown in Figure 1.20.

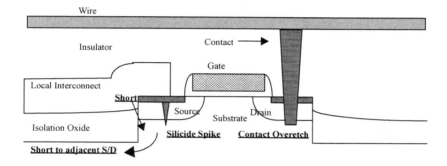

FIGURE 1.20 S/D Diffusion Leakage Mechanisms

Either mechanism reduces the effectiveness of the junction diode, allowing leakage between source or drain and the bulk or well of the device. There is little that can be done at the design level to prevent this current path from occurring.

Overetching contact openings is a common leakage path.

The opening of the contacts providing connection between the first metal wiring layer and the diffusions can also puncture the junction. In the drain diffusion shown in Figure 1.20, the etching open of the insulator separating M1 from silicon has been allowed to go too deep, and has entered the diffusion. Most processes deposit an "etch-stop" underneath the insulator. The etchant used selectively reacts with and removes the insulator at a much greater rate than the etch-stop material. Etch chemistries are available to etch silicon dioxide at rates roughly fifty times that of silicon nitride for example. A different reactive ion etch (RIE) then opens the etch-stop so that the filled contact can complete the connection. If any of the steps or films are out of tolerance, then the underlying diffusion can be disturbed as shown, and shorting from the contact to substrate can result.

It should be noted that as consequence of scaling, the diffusion-to-well diode junction is getting closer and closer to the silicon surface, as diffusion implants become shallower. Etch selectivity and control will become even more important in controlling spurious leakage sources.

Borderless contacts/local interconnects can create additional leakage mechanisms

In technologies using borderless contacts or local interconnects, conducting structures (titanium nitride and tungsten are two examples) cross the junction/isolation boundary thereby introducing yet another structure capable of introducing leakage from a diffusion to substrate or well. This is illustrated Figure 1.20. Depending on the details of the junction-isolation structure and contact process, the contact material/liner may find its way down the isolation edge allowing the conductor to contact the diffusion too close to the depletion region below. Generation currents will then be introduced into the depletion zone and become manifest as reverse-bias junction leakage.

> **Rule of thumb:** Borderless-contact-enabled shorts to the substrate or well can be on the order-of-magnitude of tens to thousands of nA per occurrence.

Various crystallographic defects in silicon can add current paths.

The layering of multiple films of various compositions, and their exposure to temperatures which exceed 800 °C create high levels of mechanical stress at the silicon surface. In addition, growth of silicon dioxide in selective areas, pushing the resolution limit of the fabricator, places stress on minimum dimension structures with the least resiliency. Oxidized silicon expands in volume, creating built-in stress. This can result in a fracture in the silicon lattice immediately under the defined structure, causing a current path from the structure to the substrate.

Another cause of fractures is incomplete annealing of damage associated with ion implantation of source/drains, extensions and even deep wells. It has been observed that with higher dopant concentrations, the annealing meant to heal silicon lattice damage can drive the imperfections to cluster along a discontinuity in the lattice, forming a "line-type defect." When the line defect approaches an implanted junction, another parasitic current path to ground or V_{DD} may be established. There is not enough data available in the literature to generalize to a rule of thumb. A TEM photo of a line defect found immediately under a gate is shown in Figure 1.21. Some silicon defects are relatively benign unless 'decorated' by metal (or other) contaminants.

Micromasking causes inadvertent printing of undesired shorts or opens.

Micromasking is shadowing of the wafer by opaque foreign material or mask defects during exposure of photoresist defining critical dimension structures. These shadows cause the printing of "bridging" shorts between wires, or breaks in connecting lines.

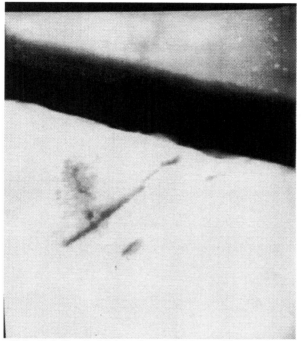

FIGURE 1.21 "Line" defect in highly doped 0.25μm silicon substrate. (Photo courtesy of C-P Eng, IBM Microelectronics, East Fishkill, NY)

Micromask defects are either repeating or random: repeating defects are typically mask problems; random defects commonly are caused by foreign material or systematic process problems. Although most are either detected during test or eliminated at Burn-in[16], some level ultimately escape to the shipped application.

---

16. Burn-in refers to the high voltage and high temperature stress testing applied to parts prior to shipment to ensure reliability. Burn-in accelerates the failure of weaknesses in interconnects and gate oxides, so that they may be screened out prior to shipment.

Process Variability   **35**

1.4 Back-End-Of-Line Variability Considerations

Until fairly recently, BEOL variation was not a primary design concern, as the percentage of total wire capacitance found in line-to-line coupling was more manageable, and static combinatorial logic was quite noise-immune. With the growing use of alternative high speed circuit topologies, and with scaled wire pitches, larger dies, and lighter relative gate loads, variation in interconnect resistance and capacitance now must be anticipated. Although BEOL processing is fairly independent of FEOL, its tolerance must be considered when defining FEOL parameters such as device threshold voltage. Noise generation is strongly dependent on small variations in wire height and spacing, as we will address in Chapter 4.

Interconnect RC Delay has dependencies on neighboring signal transitions, temporal process line drift, and on-chip spatial variation. Transient signal dependence will be explored in Chapter 4, and will be shown to have first order effects on delay and noise response. Spatial BEOL parameter variation, as with the tolerance of many FEOL parameters, comprises chip, wafer, and lot mean variation, as well as local on-chip deviations.

Two common practiced processes achieve tightly-pitched wiring levels. In *Subtractive etched processes*, a blanket layer of the conductor is electronically "sprinkled" over an insulator, then patterned by photoresist, and selectively removed by Reactive Ion Etching (RIE) to form independent wires. Insulator (usually TEOS oxide) is then applied, filling the space between wires and providing an insulating layer above the wires (See Figure 1.22).

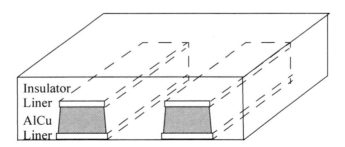

FIGURE 1.22 Sputtered-and-RIE processed interconnect

In *damascene*[17] processes, the insulator is etched rather than the metal, and then metal fills the etched areas via sputtering, electroplating, or chemical vapor deposition. The surface is then sequentially etched and polished back to where the wires are isolated by the insulator, as shown in Figure 1.23. This is the preferred process for copper metallurgies because of the difficulties in etching copper [1.20].

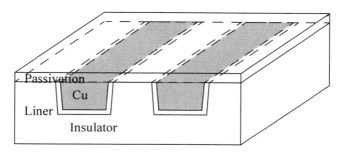

FIGURE 1.23 Damascene interconnect technology

Sources of interconnect RC delay tolerance is shown in Table 1 for a planarized BEOL using Copper in a Damascene process.:

Parameter	Primary Tolerance Driver	Affects	Impact on RC
Contact/Via Resistance	Insulator Etch Integrity	R	5%-20%
Metal Resistivity	Metal liner thickness, wire profile	R	30% - 45%
Line Width/Space	**Etch rate/pattern density**	**R, C**	**15% - 30%**
Dielectric Thickness	Deposition/polish depth	C	<5%
Dielectric Permittivity	Dielectric Stack Composition	C	<1%
Wire Thickness	**Etch Rate/Pattern Density**	**R, C**	**10%- 25%**

TABLE 1-1 Interconnect RC Dependencies. Major parameters boldfaced

From Table 1 , it is apparent that line width and space, and wire thickness account for most of the total potential **on-chip** wiring delay variation and, in fact, vary widely on a given chip. Resistivity and permittivity tend to be quite stable over a given lot of wafers The cumulative RC delay tolerance with the above considerations is shown in Figure 1.24. The reader may note that, depending on path composition and the characteristics of the process used, the BEOL potentially introduces more delay variation than the FEOL. The electrical response of this variation is explored in more detail in Chapter 4.

---

17. Damascening traces back to the process used in ancient Europe and the Middle East to inlay gold or silver into the iron of an exalted warrior's sword

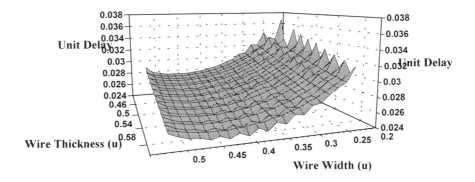

FIGURE 1.24 Wire RC Delay composite tolerance surface

1.4.1 Wire Resistance

Resistance tolerance comes from fabrication practice, materials, and wearout.

Interconnect sheet resistance bounds logic propagation speed by limiting the rate at which a capacitive load can be charged or discharged through a moderately long wire. Since that load is most likely the active gate of a sequential logic circuit, the delay required to charge that gate through its switch point becomes key to overall performance. Sheet resistance of the wire is determined by the resistivity of the materials used, and the configuration of those materials.

Interconnect resistivity, measured in μohm-cm, is a characteristic of the species of base conductor used. In practice, liner and wire thicknesses are substantial contributors to resistance variation.

Liners are thin layers of another species enclosing the conductor to block its attack by outside ions, to prevent conductor diffusion into adjoining insulator, and to reduce the likelihood of electromigration fails (see "Electromigration" below). A cross-section of a wire formed in a planarized damascene-type metallization process is shown in Figure 1.23.

During the anneal of the interconnects after initial deposition, the liner material reacts with the conductor to form an alloy. In a typical sputtered aluminum process, for example, titanium tri-aluminide is formed by reactions with the Al wire's titanium liner. Thicker liners draw more of the conductor metal into the resulting alloy. Since the alloy has higher resistivity than the core conductor, the composite wire resistivity responds measurably to changes in liner thickness.

The conductor's sheet resistance also has an inverse dependence on wire thickness, which alters the cross-sectional conduction area. Wire thickness variability is strongly process-specific, but is known to be affected by local pattern density, position on wafer, plasma etch non-uniformity, and wire width. *RIE Lag* causes wide shapes etched open by reactive-ion-etching to etch disproportionately wider and deeper, as more ions impacting from shallower angles (Angle ϕ_3 in Figure 1.25) can still partici-pate in the etch, while narrow openings require steeper angle for entry (Angle ϕ_1 in Figure 1.25).

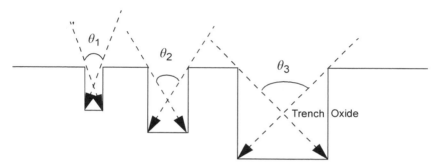

FIGURE 1.25 RIE Lag in etched oxides

Thickness variation can be controlled by constantly keeping the interconnect in the proximity of oxide. During the CMP process[18], metals, which tend to polish away more easily than oxide insulator, will form a concave region if the polishing pad is not supported periodically by oxide. Two ways of retaining sufficient oxide area is by defining a maximum wire width, or by inserting oxide "plugs" in wide wires to resist "dishing" of the conductor.

As we will explore below, wire width has a substantial impact on line capacitance as well as resistance; it is a first order process control concern.

> **Rule of thumb:** Typically, lateral sheet resistance, measured in ohms/square, has a tolerance less than or equal to +/- 10%.

Wire resistance can also increase over the life of the product due to wearout. *Elec-tromigration (EM)* is the gradual depletion of conducting metal atoms in an intercon-

---

18. Chemical-Mechanical Polishing (CMP) a technique which planarizes silicon preceding the fabrication of each interconnect layer. It combines chemical etching and mechanical polishing to remove oxide contour of underlying structure.

nect. It is facilitated by high temperature and high unidirectional current density in the wire [1.21]. Voiding of the metal is caused by collisions of conduction electrons with atoms of the conductor. These collisions liberate the atoms, which then move via diffusion along metal grain boundaries. Figure 1.26 shows a scanning-electron-microscope photograph of an electromigration casualty intentionally caused by microprocessor stress.

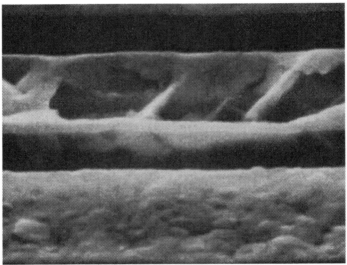

FIGURE 1.26 SEM photo of electromigration occurrence on conductor (from P. C. Li and T. K. Young , *IEEE Spectrum,* September 1996, p.75, © IEEE 1996)

> **Rule of thumb:** Although severe EM can cause virtually complete opens on aluminum wires, assume lone interconnects may degrade in resistance by up to 10% by End-of-Life for in applications with high current densities.

Because the voiding is driven by electron drift, conductors with AC current are believed to be at less risk for EM damage, but are still vulnerable to damage caused by local Joule heating.

Unlike wearout mechanisms such as hot carrier degradation, which occurs at a known rate, electromigration is a probabilistic event. The likelihood that an EM-driven failure will be observed is a function of the number of violations in the maximum allowed current in wires on a chip, the severity of the violations, the expected lifetime of the part, wire cross-sectional area, and temperature. Since current density is a function of V_{DD} and device junction temperature, the fail rate and lifetime of a product is strongly related to application use conditions. Figure 1.27 shows the electromigration-limited lifetime dependence on temperature and speed for a recent commercial microprocessor.

Electromigration can be addressed by a number of process enhancements. Aluminum-based interconnect processes introduce 1% to 5% copper to the conductor metal to form a grain size which supports less diffusion transport, much as sand and gravel form an aggregate which packs more tightly than gravel alone on unpaved roads. In aluminum metallurgy, another improvement is the use of a titanium film above and below the conductor to ensure continuity in the interconnect. Although the resulting alloy has higher resistance (see "Wire Resistance" on page 38") electromigration fails are made less likely. As technologies move to copper interconnect metallurgy, electromigration becomes less of a concern. For AlCu processes, however, the designer must closely observe current density guidelines.

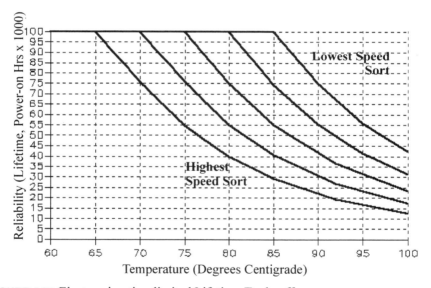

FIGURE 1.27 Electromigration-limited Lifetime Trade-offs

1.4.2 Line Width/Space

Wire width/space variation has a strong local density dependence.

Wire width and space is a primary parameter in resistance and capacitance variation. Changes in one directly come from the other at a fixed pitch. Extremes contained within the process window directly translate into delay variability. Wider wires decrease connection resistance, but increase lateral line-to-line capacitance. As shown in Figure 1.28, extremes in either direction reduce performance. While it would be attractive to center a process in the flattest portion of the process window, the resulting noise generated from line-line coupling at less space becomes expensive in the FEOL in noise margin. Since wire width and space will vary across a given chip, the design must accommodate the impact of variations in resistance and capacitance, as well as delay.

> **Rule of thumb:** Wire width variation is specific to the implemented process. None-theless, it is reasonable to expect +/- 10% variation on wire width at fixed pitch for a minimum width wire. For worst-case modeling, assume the entire variation will be seen on one chip, and half the width variation will be seen on long wires on one level.

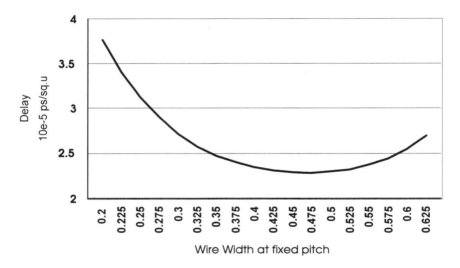

FIGURE 1.28 Interconnect Performance, Width/Space Dependence in a Cu process.

Wire width/space variation can be reduced by homogenizing as much as possible the polygon density on wiring levels, and restricting wire size. In order to maintain a tight density range, wire fill algorithms are employed after normal placement and wiring is complete to achieve uniform percentages of wire area and space, as shown in Figure 1.29.

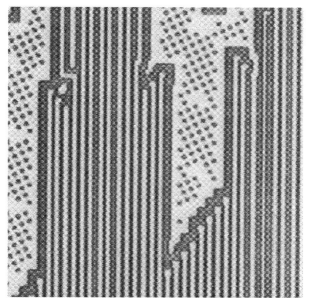

FIGURE 1.29 Metal fill (from "A 600MHz Superscalar RISC Micrprocessor with Out-of-Order Execution," Bruce Gieseke, etal, ISSCC97 FA 10.7, © IEEE 1997)

1.4.3 Dielectric Thickness and Permittivity

Vertical dielectric thickness is usually secondary to the wire's width, space, and thickness in performance impact.

The insulator between wiring levels is a film which is deposited conformally on the wafer, and enjoys high film uniformity. In a planarized process, however, variations in pattern density can cause surface concavity during CMP. The materials (metals, oxides, nitrides) all polish at different rates; variations in topography causes dielectric removal to occur at differing rates.

Dielectrics have quite stable permittivity. The use of multiple dielectric materials between two conducting layers, each with separate dielectric constants and independent thickness tolerance, introduces additional modeling and tolerance challenges. The dielectric constant of insulator between wires is very important, as it comprises the majority of wire capacitance (Figure 1.30, "Percentage Line-to-Line capacitance over multiple generations," on page 44). Since interconnect levels above and below are orthogonal, and because wire spacing tends to be tighter than vertical spacing, the vertical capacitance is less of a concern.

> **Rule of thumb:** Line-to-Line Capacitance is typically 60-80% of the total wire capacitance for a minimum wire at minimum pitch in a fully occupied wiring environment.

Finally, as the space between wires decreases, the ability to fill that space with insulator becomes more difficult. Materials and delivery systems are constantly being updated to accommodate the impact of scaling on interconnect spacing.

1.4.4 Wire Thickness

Increased interconnect thickness achieves performance at the risk of additional coupling.

Selection of interconnect film thickness probably causes as much anguish to technology developers as any FEOL device parameter does. Since fan-out delay responds so strongly to resistance, thicker wire is a quick path to performance. The pressure of scaling, however, has steadily moved wires closer and closer together; increasing wire thickness essentially increases the plate area of the lateral capacitors. As a result, the percentage of total wire capacitance found in lateral capacitance has steadily increased (see Figure 1.30).

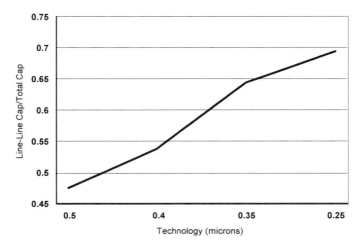

FIGURE 1.30 Percentage Line-to-Line capacitance over multiple generations

The mechanisms contributing to wire thickness variation are increasing. In RIE-etched planarized processes, dishing of the polished surface causes non-planarity on the bottom surface of the wire. Further variation is observed in the angle of the wire sidewalls due to anisotropic or excessive etching. As interconnects assume a trapezoidal profile, field lines concentrate along the lower edges, distorting line-to-line capacitance values

.In a damascene-type BEOL process, a timed etch into the insulator opens the trough to be filled with conductor. After metal fills this wire trough, the surface is CMP-planarized (see page 39); thus both upper and lower surfaces now have substantial variation contributors.

As we will see in Chapter 4, the various circuit topologies prefer specific characteristics of the interconnect structure. Die size and average interconnect lengths also help define what the BEOL should look like.

1.4.5 Contact Resistance

Contacts and vias present a challenge to scaling.

Contact and via fabrication is growing in difficulty with scaling. The contact area to the conductor below must go down by the square of the scaling factor in a given technology, yet the demand for performance has not allowed inter-layer dielectric thicknesses to fully scale accordingly. The result is large *aspect ratios* of etch depth-to-contact dimension.

There are a number of ways in which contacts and vias are formed. Generally, a hole is etched through the dielectric which insulates consecutive wiring levels, and then a liner is immediately deposited in the hole to contact the conductor below. The liner establishes the electrical integrity of the contact, blocks corrosive substances from attacking the underlying conductor, and provides a seed layer for the subsequent fill of the hole.

Common problems encountered while building contacts and vias include:

- *Achieving Ohmic Connections:* Good interface of the silicide or wire to the contact is necessary to achieve low resistance ohmic interconnects. Often Shottky-like or diode-like behavior, caused by the presence of foreign compounds or materials, is observed at the metallurgical junction, interfering with signal propagation. Ohmic connections require clean engagements of high quality liners to the contact material. Any later breach of the liner by gases used in contact deposition can cause deterioration of the underlying structure, disrupting electrical continuity.

- *Large Resistance Values:* Contacts and vias respond strongly to variations in etch, as small changes in final via dimensions change the interface area of the contact to the conductor. It has been observed that although the mean contact or via resistance can shift from lot to lot, the width and shape of the distribution about the mean remains remarkably consistent. Across a number of lots, full process window contact resistance tolerance should be expected; across a given chip, the distribution of contact resistance typically remains tight.

- *Residues:* Because the etched via or contact hole is relatively deep, it is difficult to completely remove reaction residues from the bottom of the connection. Residues, typically from etch-chemistry polymers or from gases used while depositing the via metal, can form a film between the via and the underlying metal or silicide. This film sometimes behaves as an insulator which may be "blown" open and will conduct after product "Burn-In"[19]; or it may behave as a conductor whose conductivity degrades over a period of time.

Shorts: Another result of scaling is the proximity of the contact to an adjacent gate, or the proximity of a via to a neighboring line. "Soft" short circuits can form which do not present themselves as high-current defects. The process is asked to live in a narrow window; if excessive etch time is used to ensure the cleaning out of the bottom of a contact or via, the result can be overetching of the sidewalls of the contact or via, creating wide contacts which easily shorts to neighboring gates or interconnects.

Figure 1.31 below shows a via connection from one metal layer to another, which has obvious integrity problems, yet is still functional. Even when products still are working, it is quite useful to anticipate the presence of poorly produced structures such as this via, by recognizing its signature in particularly sensitive paths during test.

Because processes vary widely in their defect modes and sensitivities, it is not appropriate to offer specific guidelines. One generalization is warranted, however.

> **Rule of thumb:** Compromised contacts and vias can range from as little as 5 ohms to many Megohms. It is reasonable to expect that contacts and vias with resistance of as large as 100 Kohms escape Burn-in Stress and are on chips that are shipped to the user. Margin should be allocated considering the probability of such a scenario in a given process.

Redundant vias and contacts are mandatory design practices which safeguard against high resistance or open connections. A single via asserts the same resistance into a net as many microns of wire: even in a robust via /contact process, automatically placing extra vias or contacts is good design practice.

---

19. A high-voltage, high temperature product stress performed to ensure high reliability.

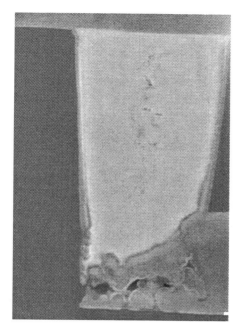

FIGURE 1.31 Resistive via (Courtesy IBM Microelectronics)

1.5 Summary

While virtually every step in the chip fabrication process introduces tolerance, a subset is capable of creating substantial changes in electrical response. These vulnerable process steps are contained in the fabrication of the device as well as the interconnect. The variations they cause in electrical parameters have both random and systematic components, and often show a layout dependence on density and orientation. In addition, one particular technology characteristic, leakage, has become much more important, with the advent of low power and dynamic circuit topologies. A collection of process mechanisms which cause leakage in particular structures was described; their collective significance will become evident in Chapter 3.

With all the performance variation that a part can see during fabrication, it sometimes seems miraculous that VLSI products function at all. What keeps CMOS practical is that most circuit design styles are quite forgiving. In addition, margin designed into products is "dynamically" re-allocated to resolve a marginal electrical parameter associated with a specific lot, wafer, or chip. Within bounds, the performance of various logical components on chip will track each other across the process window, as all on-chip components receive essentially the same mean process; clearly the design must operate across the expected process window. Of more concern is the tolerance associated with on-chip parameters which do NOT necessarily track each other or the chip mean. Budgeting cycle time for this component of delay tolerance requires experience and insight. As we will explore in Chapter 4, enough design margin must be built into the product to assure high reliability and yield, yet minimized so that the component can remain competitive in performance and price.

REFERENCES

[1.1]   L.D.Yau, "A Simple Theory to Predict Threshold Voltage of Short-Channel IGFETs," *Solid State Electronics*, 17, pg. 1058 (1974)

[1.2]   D. G. Chesebro, et al, "Overview of gate linewidth control in the manufacture of CMOS logic chips," *IBM Journal of Research and Development*, Vol. 39, No. 1/2, pp. 189-199, January, (1995)

[1.3]   A.Bryant, W.Hansch and T.Mii, "Characteristics of CMOS Device Isolation for the ULSI Age," *IEEE Technical Digest of the IEDM*, pp.671-674, (1994)

[1.4]   M.Kuhn, et al, "Ionic Contamination and Transport of Mobile Ions in MOS Structures," *J. Electrochemical Society*, Vol. 118, pp. 966-970, (1971)

[1.5]   W. Clark, et al., "Mobile Ion Collection Points in Silicided CMOS Devices," *1989 Symposium on VLSI Technology Digest of Technical Papers*, Kyoto, Japan, (1989)

[1.6]   C. Hu, et al., "Hot-Electron-Induced MOSFET Degradation-Model, Monitor, and Improvement," *IEEE Transactions on Electron Devices*, Vol ED32, No. 2, pp 375-382, February, 1985

[1.7]   J. W. Lyding, et al., "Reduction of hot electron degradation in MOS transistors by deuterium processing," *Appl. Phys. Lett.*, Vol. 68, pp. 2526-2528, 1996

[1.8]   T.Y.Huang, M.Koyanagi, A.G.Lewis, R.A.Martin and J.Y.Chen, "Eliminating spacer-induced degradations in LDD transistors," *Proceedings of the 1987 International Symposium on VLSI Technology, System and Application* (Taipei), p. 260 (1987)

[1.9]   C.E.Blat etal., "Mechanism of negative bias temperature instability," *Journal of Applied Physics*, Vol. 69, No. 31, February 1991, pp. 1712-1720

[1.10] K.Chen, C.Hu, P.Fang, M.R.Lin and D.L.Wollesen, "Predicting CMOS Speed with Gate Oxide and Voltage Scaling and Interconnect Loading Effects," *IEEE Transactions on Electron Devices,* Vol. 44, No. 11, pp 1951-1957, Nov. 1997

[1.11] C.-Y.Lu, J.M.Sung, H.C.Kirsch, S.J.Hillenius, T.E.Smith and L.Manchanda, "Anomalous C-V characteristics of implanted poly MOS structure in n^+/p^+ dual-gate CMOS technology," *IEEE Electron Device Letters*, Vol. 10, No.5, pp. 192-194 (1989)

[1.12] R.W.Mann, L.A.Clevenger and Q.Z.Hong, "Kinetic analysis of C49 TiSi2 formation at rapid thermal annealing rates," *J. Appl Phys.,* Vol. 72, no.10, pp.4978-4980 (1992)

[1.13] J.R.Brews, "Subthreshold Behavior of Uniformly and Nonuniformly Doped Long-Channel MOSFET," *IEEE Transactions on Electron Devices,* Vol ED26, No. 9, pp 1282-1291, September, 1979

[1.14] B.Pellegrini, "Reverse current-voltage characteristic of almost ideal silicon p-n junctions," *J.Appl.Phys.* 69 (2), (1991)

[1.15] T.Y.Chan, J.Chen, P.K.Ko and C.Hu, "The Impact of Gate-Induced Drain Leakage Current on MOSFET Scaling," *IEEE Technical Digest of the IEDM*, pp.718-721, (1987)

[1.16] "The National Technology Roadmap for Semiconductors Technology Needs," 1997 Edition, Semiconductor Industries Association, pp.99-113, (1997)

[1.17] J.F.Ziegler, "Terrestrial cosmic rays," IBM Journal of Research and Dev., Vol. 40, No. 1, pp.19-39, January 1996

[1.18] C.Lage, D.Burnett, T.McNelly, K.Baker, A.Bormann, D.Dreier, V.Soorholtz, "Soft Error Rate and Stored Charge Requirements in Advanced High-Density SRAMs," *IEEE Technical Digest of the IEDM*, pp.821-824, (1993)

[1.19] L.Van den Hove, R.Wolters, K.Maex, R.De Keersmaecker and G.Declerck, "The use of $CoSi_2$ as compared to $TiSi_2$ for a self-aligned silicide technology," *Symposium on VLSI Technology Digest of Technical Papers*, Kyoto, Japan, pp. 67-68 (1987)

[1.20] "The National Technology Roadmap for Semiconductors Technology Needs," 1997 Edition, Semiconductor Industries Association, p.139, (1997)

[1.21] P.-C.Li and T.K.Young, "Electromigration: the time bomb in deep-submicron ICs," *IEEE Spectrum*, pp.75-78, Sept. 1996

CHAPTER 2

Non-Clocked Logic Styles

2.1 Introduction

Non-clocked logic is ubiquitous in electronic design, due to a number of consider-ations including:

Low power consumption

Straightforward delay rule timing

Inherent reliability and noise immunity

Process variation and defect tolerance

Migratability into successive technology generations.

Deterministic diagnostic capability.

It has been said that the physical implementation of a design will never be built solely with one circuit style. While the content percentage of newer high-speed alternatives in microprocessors is steadily increasing, it is safe to say that unclocked, static cir-cuitry will be onboard CMOS designs in some quantity for the foreseeable future.

Generally, non-clocked circuit families do not require logic preconditioning, nor do they segment the microprocessor's clock cycle into distinct "evaluate" or "precharge" periods. Rather, the entire interval from when valid data appears at the output of a prior storage element to when data must be captured in a successive storage element can be occupied by sequential circuits. Non-clocked circuits respond instantaneously to changes in their inputs. If inputs to a non-clocked circuit do not change, then its outputs do not toggle, assuring that low-activity-factor paths return lower non-switching power consumption.

Although the non-clocked classes of circuits consume lower total power, propagation of logic along their path tends to charge more capacitance than other styles. The extra capacitance is associated with driving large PFETs as well as NFETs. Static power savings is realized in the lack of on-board logic clocking infrastructure needed to execute logic, and in the absence of DC current paths[1]. Transistor sizing is key in non-clocked circuits, in order to achieve performance targets. Power is minimized when circuitry utilizes the smallest possible devices, reducing the capacitive load of preceding stages.

> **Rule of thumb:** A capacitive gain of 2.7x per stage of static logic optimizes power-delay product [2.1].

Random logic libraries typically contain "books" of identical function at multiple output power levels to drive various loads. The wider output devices of the higher power books have lower "on-resistance" to V_{DD} and ground, which reduces the RC delay driving larger output capacitances. Since higher power circuits contain either more stages and/or larger input devices, they also consume more power. Hence it follows that sequences of circuits which provide output sooner than needed by using over-sized books is both unnecessary and wasteful. Further, it can contribute to "design explosion" as shown in Figure 2.1. Larger devices drive bigger die which create longer wires. The delay associated with the RC of longer wires is addressed with big-

---

1. Complementary PFET / NFET action eliminate most DC current paths, but allow a momentary current spike during transitions. See page 56.

ger devices, fueling non-convergence. Determining the fastest and most power-effi-

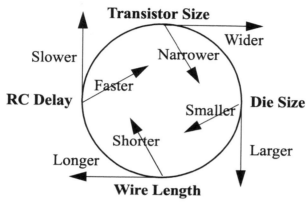

FIGURE 2.1 "Design Explosion" vs. "Design Convergence"

cient circuit implementation of complex logic functions requires extending the above discussion to the broader application.

> **Rule of thumb:** Assess the *logical effort* contained in each circuit to select proper device sizes which realize the fastest overall operation [2.2].

The delay of a non-clocked logic path has two predominant components: the intrinsic delay of the internal logic within each book, and the RC delay of the output signal driving fanout interconnect and gate capacitance through the interconnect. Since intrinsic delay may be approximated as a constant, and because delay due to output load goes as R x C, total delay can be simply represented as a function of (a) input rise time, (b) output load, plus (c) an intrinsic gate delay dependent only on processing parameters and operating conditions.

Generally speaking, non-clocked circuits are quite reliable. Nodes which determine the value of successive stages are never left to float, providing implicit noise immunity. Careful device size selection allows PFET/NFET device ratio tuning to achieve desired switch points and unity gain points for specific noise concerns. Because nodes are strongly held at all times in a complementary fashion, non-clocked CMOS is especially forgiving to defects and process variation. Since control devices receive full gate voltage even after its capacitive load has been charged, the device remains inverted, and current is readily available if needed to overcome short circuit paths or leakage mechanisms caused by minor defects. These mechanisms and their circuit sensitivities are discussed in Chapter 1. This high degree of tolerance enables the easy translation of a non-clocked design into a successive generation of the technology.

The sequential nature of non-clocked circuits supports deterministic diagnosis of fails. The state of each node is a direct consequence of the existing inputs into the circuit driving that node. A disciplined design approach allows one to work backwards through the cone of logic which is producing an incorrect output, to converge upon the node which has an unexpected short circuit or open circuit creating the fail. Although the practice has area, power, and performance costs, its advantages have been recognized in the literature [2.4].

The balance of this chapter surveys selected non-clocked static, DCVS, and pass-gate structures which have appeared in the literature. While some more recent topologies shown are clearly superior in performance, density, or power, the described predecessors provide insight into the development of non-clocked logic. These approaches efficiently execute non-clocked transitions from one machine state to another. Each design problem presents different challenges, and so the reader is encouraged to consider the trade-offs of each of these styles. While it is impractical to include all of the non-clocked circuit families which have appeared in the literature, the predominant concepts used in non-clocked families are discussed in Section 2.2, through Section 2.4.

2.2 Static CMOS Structures

Classic Static Combinatorial Logic

Pulsed Static CMOS Logic

Static Combinatorial CMOS circuitry has been the mainstay of the industry since the advent of CMOS technology. Static CMOS is a complete logic family; both true and complementary outputs of any needed logic function can be generated anywhere along the logic path. Static logic is typically propagated in one polarity[2].

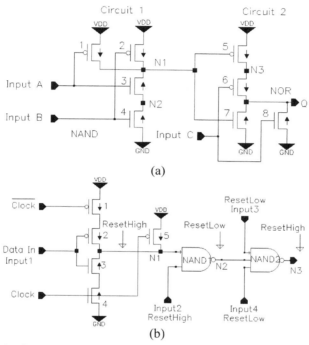

(a)

(b)

FIGURE 2.2 Static CMOS Logic styles: (a) Classic Static Combinatorial Logic; (b) Pulsed Static Logic.

2.2.1 Static Combinatorial CMOS Logic

A combinatorial static logic path executing the NAND-NOR functions is shown in Figure 2.2a.

---

2. Common referred to as *Single-Ended* logic.

Function

With input B high, transitioning high input A pulls Circuit 1(NAND) output node N1 and node N2 to ground. In the subsequent Circuit 2 (NOR), node N3 rises to voltage V_{DD}. With input C low, NOR output Q switches high, as shown in Figure 2.3.

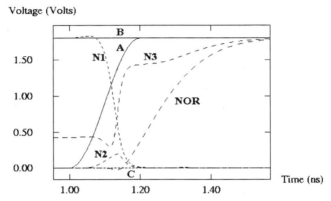

FIGURE 2.3 Static Combinatorial CMOS operation of circuit shown in Figure 2.2a.

Characteristics.

Strengths	Weaknesses
Low power	Crowbar currents
Superior yield/defect tolerance	High fan out load
✓ High test coverage/superb diagnosability	High noise generation ✓
✓ High noise immunity	

TABLE 2-1 Static Combinatorial CMOS Attributes

Static CMOS enjoys the advantages generally described in Section 1.1, but also brings noteworthy idiosyncrasies. As inputs to a static gate transition from high to low or low to high, a short circuit current, or "crowbar current" flows through both devices during the interval of the transition when both devices are on.

> **Rule of thumb:** Power consumed by crowbar current is a function of the input transition's slew rate, but consumes approximately 15% of total chip power in solely-static designs.

Lower thresholds improve performance by reducing the input voltage swing necessary to initiate stage transition. Unfortunately, as NFET and PFET device thresholds are reduced to improve performance, power lost to crow-bar current increases. Many studies of the relationship between performance and cross-over current appear in the literature [2.5].

Static combinatorial CMOS is also thoroughly testable. The output of a static circuit is always dependent on its input. By conforming to Level-Sensitive Scan Design (LSSD) convention, incorrect outputs can be traced back through the logic cone producing it, and the fault or logic error causing the incorrect output can be deterministically identified [2.4].

2.2.2 Pulsed Static Logic (PS-CMOS)

We've just begun our exploration of circuit styles, and already the author is disrupting this chapter's structure by introducing Pulsed Static CMOS, a *clocked* style! Pulsed Static logic, however, propagates logic through preconditioned, but still unclocked, circuits [2.3]. A logic path implemented in Pulsed Static CMOS is shown in Figure 2.2b.

Function

Assume, initially, Clock is low, and $\overline{\text{Clock}}$ is high. Preconditioning PFET device 5 is on, and devices 1 and 4 are off, driving node N1 high. With Input2 coming from a circuit also resetting high, node N2 resets in the low state. Similarly, with Input3 and Input4 reset in the low state, node N3 resets in the high state. As Clock transitions high, devices 1 and 4 turn on, and device 5 turns off, driving the state of node N1 to assume the inverted state of Input1. Node N1's state then propagates through the remainder of the logic chain in a sequential manner.

Characteristics

Strengths	Weaknesses
Low power	Maximum clock speed limited by reset
Higher static performance	Cumbersome monotonic stage operation
No dynamic noise immunity problems	Restricted connectivity
High static circuit testability	Complex clocking

TABLE 2-2 PS-CMOS Attributes

As the original patent in the reference points out, pulsed static logic avoids three disappointing attributes of the static circuit style discussed in Section 2.2.1;

1. Delay dependence on previous logic input patterns,

2. False transitions caused by intermediate states of inputs, and

3. High fanout load caused by PFET And NFET structures in next stage.

Because static logic output is always deterministically defined by its input, the creators of PS-CMOS observed that if the initial input signal out of a latch is timed and monotonic in its transitions, then so will follow the rest of the path. Once the evaluate direction is known, then devices can be optimized for the critical path they are in. This is achieved by clocking two patterns out of the latch, which propagate sequentially through the logic; first a signal resetting the path into a known state, and then the actual logical input for evaluation. Further, since specific static circuits have slower transitions in one direction, their reset is chosen in that direction. Superb noise immunity, and dramatic reduction in clock load compared to the dynamic embodiment are achieved by eliminating dynamic floating nodes. In addition, fan-out load and power is greatly reduced.

When we later study clocked, alternating polarity structures[3], similarities to PS-CMOS will be apparent. Both circuit families require attention to the precharge/precondition polarity of a given stage's logical inputs. All preconditioning, however, usually occurs in parallel in clocked logic paths; in PS-CMOS, it must propagate through the logic path. In the fastest designs, the highest clock speed will be limited by preconditioning as well as evaluation. In addition, the design must provide for filler logic to accommodate path delay differences of multiple paths arriving at the target latch inputs simultaneously. Sufficient margin must be allowed to guarantee the slower reset "wave" to complete before the faster evaluate "wave" can catch it.

2.3 DCVS Logic

Differential Cascode Voltage Switch Logic (DCVSL)

Differential Split-Level Logic (DSL)

Cascode Non-Threshold Logic (CNTL)

Differential Cascode Voltage-Switched Logic (DCVSL), first described at the International Solid-State Circuits Conference in 1984 by L.G. Heller, etal. [2.6], is the foundation of many contemporary higher-speed circuit structures. Inspired by Domino CMOS, DCVS influenced most of the subsequent differential innovations described in this text; the DCVS ISSCC paper is cited in a majority of chapter 2 and 3 references. The authors have been privileged to work with Dr. Heller and explore this important concept. His contributions to CMOS digital logic are widely acknowledged.

---

3. See Section 3.3, Alternating-Polarity Domino Approaches on page 107.

There are a number of DCVS variations in the literature, all employing the basic con-cept of using pairs of *differential*[4] logic inputs to flip a static cross-coupled device pair and store an output state. Logical arrangements, or *trees,* of stacked evaluation devices potentially couple the circuit's output node to ground (or to a developed sig-nal or V_{DD} in some structures), conditional on the result of the evaluation.

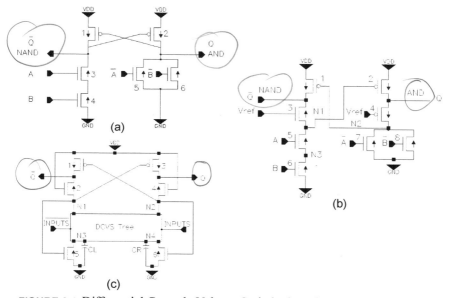

FIGURE 2.4 Differential Cascode Voltage-Switched Logic Structures: (a) Basic DCVSL AND Gate with PMOS Cross-Coupled Loads; (b) Differential Split Level Logic AND gate; (c) Cascode Non-Threshold Logic.

2.3.1 Differential Cascode Voltage-Switched Logic (DCVSL)

The DCVS implementation of a AND/NAND is shown in Figure 2.4a.

---

4. Differential circuits require logic structures whose evaluated result on a given side of the logic design is exclusively the complement of the evaluated result of other. For example, in Figure 2.4a, the left side of the DCVS structure shown produces the **NAND** and the right side the **AND** of inputs **A** and **B**.

Function

Referring to Figure 2.5, with input B high and input $\overline{B}$ low, transitioning high input A and transitioning low input $\overline{A}$ are fed to the differential evaluate tree, latching node $\overline{Q}$ low, and node Q high.

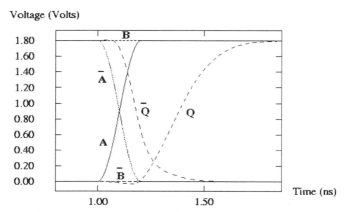

FIGURE 2.5 DCVSL operation of circuit in Figure 2.4a

Characteristics

Strengths	Weaknesses
Superior logic density	Dual rail wiring
Implicit invert available	Struggle between PFET output strength and latch hysteresis
High reliability and noise immunity	Higher device count in some applications

TABLE 2-3 DCVSL Attributes

The advantage of Differential Cascode Voltage Switching is realized in the logic density achieved by evaluating complex trees of logic in one delay stage. Differential-pair logic trees may easily be 4 devices tall, and process (2^4-1) inputs. Further efficiency is achieved in the elimination of large PFETS from each logic function executed in the tree. Boolean functions are implemented in NFETs only; the PFETs serve solely as pull-up devices. Differential Static Cascode-Voltage Switched logic offers implicit noise immunity at each stage, due to its cross-coupled nature. Automated logic tree synthesis has been shown to be easily accomplished with existing design tools [2.7].

Performance is somewhat compromised by the hysteresis associated with toggling the load devices. The PFET load devices must be small enough that its on-current is easily overcome by the switching logical pulldowns, but large enough to drive high outputs with acceptable delays. Normal process tolerance can make it hard to remain within a window of acceptable device strengths. Until the output nodes flip, the evaluate tree sinks load current, highlighting the power trade-off that cross-coupled load devices require.

2.3.2 Differential Split Level Logic (DSL)

Differential Split Level Logic (DSL) addressed the problem in DCVS that until a given stage flips, there will be substantial current to ground [2.8]. The DSL solution to reducing this current spike also reduces signal voltage swing, realizing a consequent active power reduction. Static power may increase in some implementations, however, due to a low current DC path present. A simple DSL AND/NAND circuit is shown in Figure 2.4b.

Function

With input B high and input $\overline{B}$ low, assume input A transitions from a low (at ground) to a high (at voltage V_{DD}), and input $\overline{A}$ transitions from a high value to a low value, as shown in Figure 2.6. With bias V_{REF} set at ($V_{DD}/2+V_{TN}$), PFET load device 1, connected via node N2 as shown, receives a bias of $V_{DD}/2$; PFET load device 2 has its gate grounded through devices 5 and 6. Signal $\overline{Q}$ develops a low output (ground) through devices 3-6, and signal Q is developed as a high output (V_{DD}), as devices 7 and 8 are off.

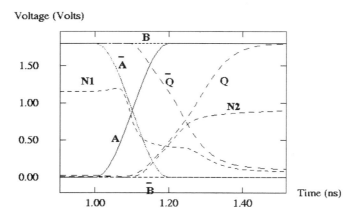

FIGURE 2.6 DSL operation of circuit shown in Figure 2.4b.

Characteristics

Strengths	Weaknesses
Lower AC power	Higher standby current, power dissipation
Superior static performance	Reference voltage generation required
High reliability	Higher device count
Reduced channel hot-electron wearout	

TABLE 2-4 DSL Attributes

DSL leverages the presence of a precision bias. Sampling output off immediately below the load PFET device provides a strong high output at full V_{DD}; a low output, however, is held at ground through the resistance of the logic tree. The original litera- ture recommends taking outputs from these uppermost nodes. Transitions are has- tened because node N1 need only to be discharged from $V_{DD}/2$ rather than V_{DD}.

The tree device gated by V_{REF} makes DSL substantially different from its DCVS ori- gin. Since the NFET pull down device never sees drain voltages higher than $V_{DD}/2$, the channel hot electron-based performance degradation is minimized. It can be argued that this enables usage of technology with channels shorter than what would be expected for the given V_{DD}, realizing higher performance. The added resistance of the reference device necessarily reduces the total possible tree height.

The presence of that device may also never allows the PFET load device on the "0" side of the output to be completely turned off, as its gate rise to $(V_{DD}-V_T)$. The presence of resulting current causes a "low" output to hover 100-200 mV above ground, causing DC power consumption. This increase, and the higher transistor count caused by the reduced logic content must be weighed against the AC power savings realized by the limited swing.

2.3.3 Cascode Non-Threshold Logic (CNTL)

As we noted in "Differential Cascode Voltage-Switched Logic (DCVSL)" on page 59, considerable power is consumed in cross-over current while switching the contents of the circuit. Additional power is consumed in achieving the full swing of the internal node. DSL reduced active power consumption by reducing voltage swing. Cascode Non-Threshold Logic (CNTL) further improves the concept [2.10]. A generic representation of CNTL Logic is shown in Figure 2.4c.

Function

True and complement logic inputs are provided to the DCVS tree of NFET devices. The load circuit comprised of devices 1-4 latches in the direction determined by the logic tree. NFET devices 2 and 4 act to limit the *upper* end of the logic voltage swing. NFET devices 5 and 6 generate negative feedback to their respective trees by increasing their on-resistance as the tree discharges, limiting the *lower* end of the logic voltage swing. Capacitors C1 and C2 are present to shunt the effect of devices 5 and 6 and moderate the negative feedback. The transfer function associated with this style has no distinct "break" at a logic threshold, but produces logically valid output levels.

Characteristics

Strengths	Weaknesses
Low power	Separate rail-level interface required
Heavy load drive capability	Compromised performance
Higher logic content per stage	Substantial area increase
Differential noise immunity	Shunting capacitors needed

TABLE 2-5 CNTL Attributes

Cascode Nonthreshold Logic makes use of a concept first implemented in bipolar transistors; by resistively dividing the power supply voltage, the required output voltage swing can be reduced, improving performance. CNTL is developed from CMOS Nonthreshold Logic (NTL), a single-ended CVS structure not presented here due to its excessive DC power consumption. CNTL improves upon DSL (See "Differential

Split Level Logic (DSL)" on page 61.) by not requiring a precision voltage reference. The devices which add additional voltage drop have their gates tied to V_{DD} or to a feedback node, instead of a reference. The negative feedback substantially reduces the power consumption of the NTL origin, but at the expense of some performance. The performance penalty is addressed with the shunt capacitor. Like other CVS-style designs, the circuit comprises a differential logic tree of NFETs, and a PFET load circuit.

As discussed in the final chapter, device threshold has failed to scale well with supply voltage. As technologies continue to reduce V_{DD} with scaling, the reduction of operating swing will become increasingly difficult.

2.3.4 DCVS Circuit Family Process Sensitivities

The DCVS circuit family shares the sensitivity of static circuitry to device transconductance. Specifically, since the evaluate path commonly comprises NFET devices, changes in NFET device channel length, threshold, or gate oxide will be immediately reflected in circuit performance. DCVS device sensitivities are briefly listed below, and are explored in more detail in Chapter 4.

1. A subset of DCVS styles, such as DSL, depend on precision reference bias voltages which anticipate specific device threshold voltages. Threshold voltage variation, therefore, has a first-order influence on referenced topographies.

2. The performance of the stacked devices found in static DCVS evaluate trees suffer from body effect[5]. The logic evaluation trees must provide low resistance paths to ground in order to switch the load circuit; positive NFET source-substrate bias voltage increases threshold voltage and introduces additional delay. Changes in device substrate sensitivity will cause similar circuit response.

3. The performance of DCVS styles which incorporate cross-coupled latches are strongly influenced by changes in PFET-to-NFET channel length and threshold voltage tracking. Changes in the relationship of PFET to NFET strength impacts hysteresis by altering the latch's V_{OUT}/V_{IN} relationship.

4. Process variations which increase capacitance, such as changes in source-drain capacitance or overlap capacitance can add tree transition delay variability, as loads must first be satisfied before the outputs can be toggled. Insuring that layouts minimize junction area and tapering tree device widths [2.11] are two techniques that can help reduce delay.

---

5. See 1.2.9, Body Effect on page 21

5. Positioning of the signal inputs on the evaluate tree based on arrival time is very powerful. Early-arriving signals should be positioned low on the tree, to discharge as much tree capacitance as possible without waiting for the last arriving input to need to do so.

6. Generally, narrow channel effect[6] is not a concern in DCVS families, as most of the unclocked styles must use wider devices to handle succeeding fan-out load or to gain the switch point of the latch/buffer load circuit.

7. DCVS topologies which reduce logic voltage swings to reduce active power or delay are especially sensitive to parametric shifts causing resistive voltage drop. As output swing is reduced, the difference in voltage between a stage's lowest possible "high" output and a following stage's lowest acceptable "high" input [7] collapses, jeopardizing noise immunity. For the same reason supply tolerance also becomes an issue[8]. The value of differential logic is significant in reduced voltage swing schemes.

2.4 Non-Clocked Pass-Gate Families

CMOS Pass Gate and Transmission Gate Logic

DCVS Logic with the Pass Gate

Complementary Pass Gate Logic

Swing Restored Pass Gate Logic

Energy-Economized Pass Transistor Logic

Push-Pull Pass Transistor Logic

LEAN-integrated Pass Gate Logic

Double Pass-Transistor Logic

FET devices configured as pass gates appeared historically as an integral component of one of the first important MOS device implementations, the one-device DRAM cell. Pass gates are also indispensable in SRAM memory. It is no surprise then, that their versatility is exploited in CMOS logic.

---

6. See Section 1.2.3, "Channel Width Effects" on page 11.

7. See Section 4.5, "Noise" on page 157.

8. See Section 4.4.1, "VDD Tolerance Effects" on page 153.

Non-Clocked Logic Styles   **65**

Pass gate logic has earned the reputation of being a "risky" design styles, and with good reason. Compared to static logic, pass gate structures carry more parametric sensitivities and fail modes. Referring to Figure 2.7, there are a number of issues which

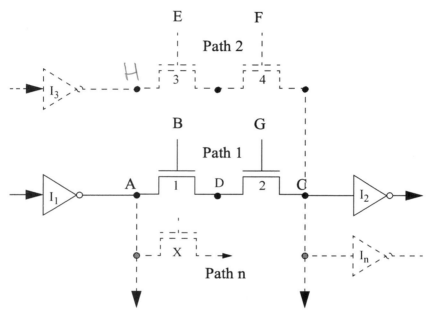

FIGURE 2.7 Pass Gate Concerns

must be addressed. The reader should note that the vulnerabilities of non-clocked pass gates are yet only a subset of those associated with the clocked version of this circuit family. See Section 3.6 on page 124 for a summary of clocked dynamic pass gate design challenges.

1. Limited fan-in capability

For a pass gate with fixed transistor sizes, the amount of current to be discharged to ground through the structure must be limited in order to achieve acceptable down levels with reasonable delay. In some configurations, a pass gate circuit may be expected to provide a level to multiple output nodes. In Figure 2.7, the cumulative gate capacitance of n output buffers, found on node C, must be discharged to ground in a timely manner through Path 1 when devices 1 and 2 are logically gated on.

2. Excessive Fan-out

The output inverter providing logic signals to subsequent pass gates must be sized to provide sufficient high and low levels to all the paths it serves. If this fan-out is excessive, the inverting buffer can be overwhelmed. In Figure 2.7 for example, Node A is tied to n logic paths; inverting buffer I_1 must be sized to deliver timely outputs with acceptable levels when coupled to additional paths, represented by device X in the illustration.

3. Noise

Unlike static combinatorial CMOS, noise on signal lines caused by interconnect coupling can be propagated through a pass-gate circuit via device gates *or* sources, node A or B in Figure 2.7. This consideration doubles the opportunity for noise-driven disturbances of the output data. Device 1 in Figure 2.7, for example, can be switched momentarily to its "on" state if

- node B couples up to a voltage *above* device 1's threshold voltage, or if
- node A couples downs to a voltage *below* device 1's threshold voltage.

If input G is high, this event appears on the input of inverter I_2, temporarily disrupting the output. Noise in static logic can not be disregarded. Noise is explored in more detail in Section 4.5 on page 157.

4. Rail-to-Rail[9] Voltage

Poor up or down supply levels can create voltage differentials on inputs. For example, when logic inputs A and B are low in Figure 2.7, they should both be at ground potential and device 1 should be firmly cut off. If the voltage level of input B drifts above ground with input G high, device 1 is more at risk for turning on and disrupting the output. At the very least, bias reduces overdrive, decreases noise immunity, and aggravates the known body effect in the pass gate structure.

5. Violation of decode exclusivity

Power, delay, and design robustness are optimized when only one path to a source signal exists at any given time, including during transient intervals. Consider, for example, the following scenario:

- devices 1, 2, 3, and 4 are all temporarily on for a brief time,
- node H is being driven to a high state by I_3, and

---

9. The voltage boundaries of a signal which moves from ground to V_{DD}.

Non-Clocked Logic Styles

- node A is driven to a low state by I_1.

DC current then briefly exists through paths 1 and 2, impacting power if occurring chip-wide, and node C is driven to an intermediate voltage, which can increase delay.

Often the path is a multiplexer input to a latch, where NFET device 2 is the clocked latch input postage, and inverter I_2 forms a latch node with an additional weak feedback inverter not shown. False inputs can cause fails by occurring before the clock on node G goes low. The weak feedback is insufficient to restore node C to the correct level after the overlapped decoding ends and before G goes low.

6. Full High Level Restoration

The exclusive use of NFET devices in the pass-gate logic path, as shown by devices 1 and 2 in Figure 2.7, causes logic "high" levels of V_{DD}-V_T on node C, delivered to inverter I_2. This voltage, presented to the output buffer (I_2 in this example), causes its PFET device to not turn fully off, allowing leakage current and loss of noise immunity.

7. Body Effect

During discharge of node C, nodes A and D will be above ground potential, due to the voltage drop across inverter I_1's pulldown device, and devices 1 and 2. As we saw in Section 1.2.9 on page 21, even transient positive voltage from source to substrate can impact performance by elevating threshold voltages. In this example the threshold voltages of devices 1 and 2 will be increased, increasing the time needed to discharge node C.

Despite the limitations described above, pass gate topologies are indeed very fast and, if used prudently, can substantially improve critical path delay. Pass gate studies in the literature find, at minimum, a 20% performance improvement over static logic [2.12]. Lets proceed, then, to examine how a number of pass-gate structures develop logical outputs.

2.4.1 CMOS Pass Gate (PG) and Transmission Gate (TG) Logic

Pass gates and transmission gates[10] are versatile circuit elements which have been used in conjunction with static CMOS for many years [2.13]. The two styles can be quite compatible with static CMOS. Although realizable in static combinatorial

---

10. Transmission gates are also known as *transfer gates*.

CMOS circuits, a function built with pass or transmission gates can substantially improve performance. The simple CMOS transmission gate stage, comprising PFET and NFET transfer devices, is shown in Figure 2.8. By omitting inverter I1 and PFET

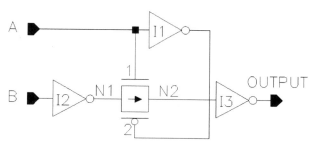

FIGURE 2.8 CMOS Transmission Gate

device 2, the circuit is referred to as a *pass-gate*.

Function

Gating logic input A going high turns on NFET device 1. Inverter I1 generates the complement of the signal which turns on PFET device 2. With both an NFET and PFET turned on, a rail-to-rail voltage swing can be propagated from the output of inverter I2 to the input of inverter I3. If input B is at ground, then node N1 is driven to

V_{DD}, and node N2 can also reach V_{DD}, driving the output to ground. Use of inverter I1 and PFET device 2 is optional, but their omission reduces the voltage of a high input to Inverter I3 to $V_{DD}-V_T$. Both cases are shown in Figure 2.9.

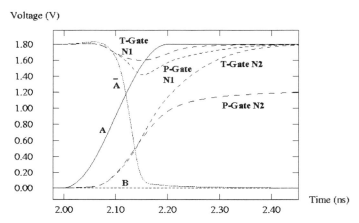

FIGURE 2.9 CMOS Transmission-gate and Pass-gate operation of circuit in Figure 2.8.

Characteristics

Strengths	Weaknesses
High speed operation	Body Effect in some topographies
Low area (pass gates only)	Source follower action in some topologies
Low power	Limited logic depth

TABLE 2-6 CMOS Pass Gate and Transmission Gate Attributes

The NFET-PFET pass gate pair forming the CMOS Transmission gate provides a more reliable means of propagating levels without additional buffer delay. The achievement of full rail-to-rail swing on the output node eliminates the cross-over leakage on the next rebuffering stage, and improves noise immunity of the circuit. By maximizing the difference in voltage between the least positive up level (LPUL) of the output and the lowest voltage still interpreted as a high by the next stage, noise margin is improved. The cost of this margin is the additional delay driving the PFET device. It is common practice to size the NFET to pass the majority of the signal's transfer charge. Often the PFET device is, in fact, eliminated, and the structure reduces to the simple NFET device pass gate. Transmission gates or pass gates are often serially connected in a path. Discrete buffering, as shown by inverters I_2 and I_3 in Figure 2.8 is also often omitted when interconnect lengths are kept short or loads kept light; the pass gate signals are then passed directly to the next logic stage.

2.4.2 DCVS Logic with the Pass Gate (DCVSPG)

DCVS logic with the pass gate is a means of extending the performance benefits associated with DCVSL into pass-gate topologies [2.14]. A DCVSPG configuration performing the XOR function is shown in Figure 2.10.

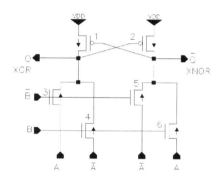

FIGURE 2.10 DCVSPG Logic implementation of logical XOR

Function

True and complement versions of logical inputs A and B are supplied to the DCVSPG circuit. If A and B both transition high, then NFET devices 4 and 6 are turned on, and NFET devices 3 and 5 are in their off-state. NFET device 4 couples output Q to $\overline{A}$, pulling Q low. NFET device 6 couples output $\overline{Q}$ to A, pulling $\overline{Q}$ up to voltage V_{DD}-V_T. PFET device 2 is responsible for developing the full voltage V_{DD} on output $\overline{Q}$.

Characteristics

Strengths	Weaknesses
Full-swing pass-gate action	Limited logical depth and load drive
Differential noise immunity	Body effect
Substantial area, power reduction	Source follower action
DCVS Floating node problem eliminated	

TABLE 2-7 DCVSPG Attributes

DCVSPG logic behaves very differently from it's static DCVS parents[11]. In DCVSPG, both the NFET and PFET contribute to pull-up performance, and both true and complement outputs are actively driven to their logical value [2.14]. At no time is either node temporarily floating, as is the case in its DCVS predecessor. The PFET device sizing sensitivity problem in conventional DCVS is also eliminated; improperly-sized PFETs do not affect functionality. These PFET devices impact only performance, by adding drain and gate capacitive load. As implied by Figure 2.11, an optimized PFET device width will maximize pull-up performance while minimizing it's pulldown performance impact.

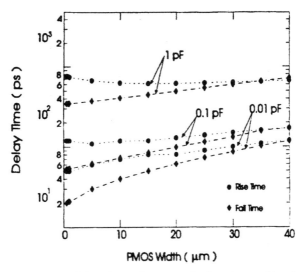

FIGURE 2.11 DCVSPG delay dependence on device size and load, with constant NFET device sizes.(F-S Lei, W Hwang, IEEE JSSC, 4/97, p. 565,© IEEE 1997)

11. See 2.3.1, Differential Cascode Voltage-Switched Logic (DCVSL) on page 59

Relief from the device size ratio concern in DCVSPG also improves power consumption, as the switchpoint for the circuit is achieved without having to first sink opposing current. Since the logic levels assert their own drive to the output nodes, hysteresis delay is eliminated; logical inputs also directly drive the gate capacitance of the cross-coupled PFETs. The PFETs restore outputs to full high levels, providing some measure of noise immunity.

2.4.3 Complementary Pass Gate Logic (CPL)

Despite continued innovations using other structures, Complementary Pass Gate Logic remains one of the simplest, fastest, and most frugal of the circuit families using transistors in pass-gate configurations [2.15]. Two CPL implementations of an XOR are shown in Figure 2.12a and Figure 2.12b.

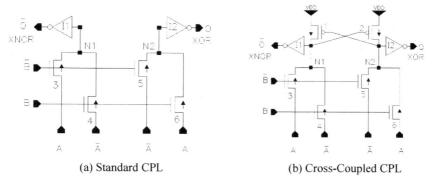

(a) Standard CPL   (b) Cross-Coupled CPL

FIGURE 2.12 Complementary Pass Gate Logic implementations of XOR

Function

Referring to Figure 2.12b, logical control inputs B and $\overline{B}$ gate A and $\overline{A}$ into differential nodes N1 and N2. Optional cross-coupled PFET devices 1 and 2 pull N1 or N2, whichever is high, from V_{DD}-V_T the rest of the way up to V_{DD}. Inverting buffers I1 and I2, tied to N1 and N2 respectively, develop outputs $\overline{Q}$ and Q. See Figure 2.13.

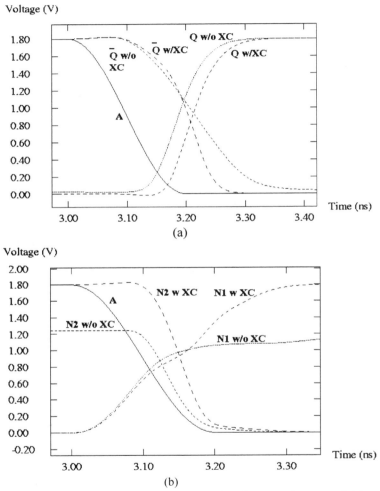

FIGURE 2.13 CPL operation of circuits shown in Figure 2.12; external (a) and internal (b) node response

Characteristics

Strengths	Weaknesses
High speed operation	Body effect
Less stages of delay, and less delay/stage	Limited logical depth
Small tree devices + NFET transconductance	High leakage power when not cross-coupled

TABLE 2-8 CPL attributes

Basic CPL logic is very fast, making use of NFET device logic trees which couple logical inputs to inverting buffers driving outputs. *In its simplest form, no PFET devices are present to complete the swing of rising inputs to the inverting buffers,* such as PFET devices 1 and 2 in Figure 2.12b[12]. The swing of nodes N1 and N2 in Figure 2.12a are then limited to a high level of V_{DD}-V_T. The inverting output buffers of each CPL circuit perform two important services:

1. Amplification of high tree outputs from V_{DD}-V_T up to full rail voltage, and

2. Capacitive output load handling capability.

For one of the two inverting output buffers in Figure 2.12a receiving gate drive of V_{DD}-V_T, its PFET is never entirely shut off, resulting in "crowbar" current and power consumption in the inverter producing the low output side. Three alternatives proposed in the literature address this crowbar current and are described below.

1. The PFET cross-coupled load structure shown in Figure 2.12b completes the transition to a full up level. This latch structure, however, reduces performance by adding extra evaluate capacitance.

2. Reducing the threshold of NFET devices 3-6 in Figure 2.12, to zero or near-zero voltage brings the high evaluate node closer to V_{DD}, but increases process complexity and noise sensitivity.

---

12. The reader may notice the similarity between cross-coupled CPL and DCVS-PG. The presence of the inverter/buffer in CPL removes the load-driving responsibility from the cross-coupled pair, but adds one inversion delay to the output.

Non-Clocked Logic Styles

3. A modified CPL driver includes feedback to restore full voltage swing. The feed-back is independent of the output load [2.17]. Because it's response time does not depend on the load's slew rate, it can respond more quickly, and effectively reduces power. The feedback-enabled CPL driver/buffer is shown in Figure 2.14 below.

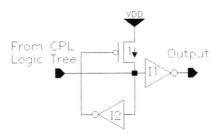

FIGURE 2.14 CPL inverter/buffer with fast-feedback

It is perhaps obvious by now that in CPL logic path, the only component *generating* high and low levels is the output inverting buffer. The NFET device logic tree only *passes* buffered CPL output signals to subsequent inverting buffers at the other end of the tree, placing signal integrity responsibility squarely on the buffer's design. The NFET devices in the evaluate tree do not need to be very wide, as their function is to provide a (dis)charge path for the load capacitance only. The logic tree output is dif-ferential, and there is no hysteresis which the pull-down must overcome.

2.4.4 Swing-Restored Pass Gate Logic (SRPL)

A number of pass-gate-based logic families use differential pass gate trees, similar to the CPL structure, to execute a required logic function. They vary, however, in their load circuit structure, attempting to improve upon CPL's power-delay product. Swing Restored Pass Gate Logic replaces CPL's cross-coupled PFET devices with a 4-

device cross-coupled latch [2.16]. A Swing-Restored Pass Gate implementation of the XOR function is shown in Figure 2.15.

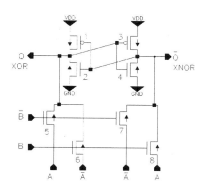

FIGURE 2.15 Swing-Restored Pass Gate Logic implementation of XOR

Function

Logical input B and its complement $\overline{B}$ drives NFET gates in a pass-transistor logic tree, leading from logical input A and its complement $\overline{A}$ to differential outputs Q and $\overline{Q}$. The latch formed by devices 1, 2, 3, and 4 completes the switching, or *swing*, of the circuit output by *restoring* full rail voltage for high levels. The latch also assists in *discharging* outputs to low levels after the tree has flipped the latch.

Characteristics

Strengths	Weaknesses
Low standby power consumption	Limited output drive restricts loads
High design margin, process tolerance	Potential for high cross-over current
Less delay via sense-amp action	Trade-off of drive strength vs. performance
More logic depth possible	

TABLE 2-9 SRPL Attributes

SRPL logic has the ability to implement any boolean logic function in a tree which is restored to full swing by the latch. Since the pull down tree must overcome the latch's PFET current besides load current, a trade-off exists between the performance of the transition and the maximum load which each stage may drive. The latch's NFET devices must be sized small enough to minimize capacitive loading and hysteresis delay, but large enough to help achieve and maintain the latch state. Ideally, once the logic tree accomplishes the flipping of the latch, the single latch NFET to ground takes over the business of producing strong high and low levels.

Compared to CPL, the logic tree's NFET devices must be substantially larger. The string of tree NFET devices in series must be able to overcome the state stored in the latch, requiring that the tree's path to ground defeat the latch PFET device, to pull the latch's intermediate node below its switch point.

2.4.5 Energy-Economized Pass Transistor Logic (EEPL)

Another improvement to CPL attempts to restore full voltage level swing while avoiding the FET horsepower necessary to overcome the hysteresis of a latch. The Energy Economized Pass-Transistor Logic [2.19] implementation of the XOR function is shown in Figure 2.16a.

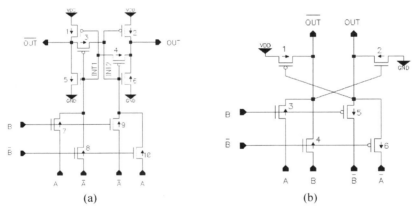

(a)   (b)

FIGURE 2.16 More recently published cross-coupled pass-gate structures; Energy Economized Pass Transistor Logic implementation of XOR (a); Push-pull Pass Transistor Logic implementation of AND/NAND

Function

Referring to Figure 2.16a, assume a difficult transition for EEPL, when logic input A transitions low, and $\overline{A}$ transitions high, while logic input B remains high and $\overline{B}$ remains low. Node INT2 is then driven to V_{DD}-V_T through NFET device 9. NFET device 6 is turned on, PFET device 2 is essentially off, and OUT is pulled low, as shown in Figure 2.17. PFET device 4 is off. In the left-sided tree, NFET device 7 is turned on, coupling A, transitioning low, to node INT1. NFET device 5 is turned off,

PFET device 1 is turned on, and $\overline{OUT}$ goes high. PFET device 3 is turned on, which couples node INT2 to $\overline{OUT}$, swinging INT2 to the complete V_{DD} rail voltage.

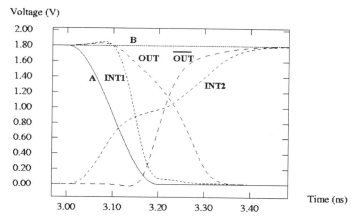

FIGURE 2.17 EEPL operation of circuit shown in Figure 2.16a.

Characteristics

Strengths	Weaknesses
Positive regenerative feedback	Lower performance at larger stage counts
Extremely low power	Limited logical depth
Low device count	

TABLE 2-10 EEPL Attributes

EEPL reduces power consumption and delay by interrupting the feedback of the latch forming the load circuit in the structure, allowing reduction in the width of the NFET devices comprising the evaluate tree. This device width reduction further contributes to the power reduction. The circuit action simultaneously provides regenerative positive feedback, providing shorter delays than comparative CPL circuits. EEPL will be a valuable logic element in low power applications where performance is still essential.

2.4.6 Push-pull Pass transistor Logic (PPL)

The innovations described in Sections 2.4.4 and 2.4.5 address improved means of recovering full V_{DD} levels while still exploiting the speed of differential pass gate trees. Both styles, however, still must include PFET and NFET devices in the load circuit to output the full rail voltages of either output signal polarity. Independent of topology, this additional capacitance reduces performance. By making the differential pass-gate trees complementary in their device usage, Push-pull Pass-transistor Logic [2.20] (PPL) disadvantages only one signal polarity on either side. The PPL structure accomplishing the XOR function is shown in Figure 2.16b.

Function

Assume logic input B transitions high and $\overline{B}$ transitions low while logic input A remains high and $\overline{A}$ remains low, as shown in Figure 2.18. In the left-handed NFET device pass gate tree, device 3 is on and device 4 is off, passing the high state of logical input A to node $\overline{OUT}$. In the right-handed PFET device pass gate tree, device 5 is off and device 6 is on, passing the input $\overline{A}$ low state to node OUT. With $\overline{OUT}$ at voltage V_{DD}-V_{TN} and OUT at ground+V_{TP}, PFET device 1 and NFET device 2 both turn on, assuring the transition of the OUTN and OUT to full rail voltages V_{DD} and ground, respectively. Evidence of this action is seen in the inflection of the output curves, occurring at the turn-on of assist devices 1 and 2.

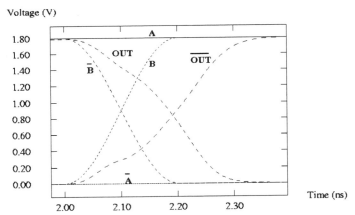

FIGURE 2.18 PPL operation of circuit shown in Figure 2.16b.

Characteristics

Strengths	Weaknesses
Push-Pull action, no need for rebuffering	Limited fan out capability
Hysteresis delay elimination	PFET drive dependence
low power	limited sequential application usage

TABLE 2-11 PPL Attributes

PPL achieves a performance advantage using complementary logic trees, and cross-coupling to pull up or pulldown devices which recover full rail voltage. By doing so, total device count, load circuit capacitance, and power consumption is reduced. The topology also effectively removes hysteresis associated with the latching action. The trade-off this style makes, however, is the use of PFET devices in the logic path, which can reduce performance: the original literature demonstrates that with limited fan-out capacitance, PFET devices in the evaluate path do not significantly impact performance or power.

The designer must also examine the paths back to V_{DD} and ground in a typical use. For PPL, V_{DD} and ground is provided by:

1. the PFET pull-up and the NFET pull-down restoration devices (devices 1 and 2 in Figure 2.16b)

2. the drivers providing logical inputs to the sources of the logic tree devices (inputs A and $\overline{A}$ to devices 3 through 6 in Figure 2.16b).

One concern must be considered. The inputs to each PPL circuit in a PPL path are provided by the output of a preceding circuit. If the entire path assumes circuit states which do not turn on the level-restoration devices (a probability of 50% per stage), the entire path refers back to the drivers of the first stage's input, resulting in potentially unacceptable voltage levels or performance.

2.4.7 Single-Ended Pass-Gate Logic (LEAP)

"Top-Down Pass-Transistor Logic" Design, promoted in the literature by a number of articles [2.17] [2.18], proposes reduction in size of a random logic library through the use of a "lean" set of simple, flexible pass gates which could be used bidirectionally, and whose inputs could be connected to either supply rail if needed for a particular logic voltage level. LEAP concentrates on synthesizing the function of full logic blocks rather than individual logic functions. Lean integration using pass gates ("LEAP" structure) of a logical 3-way AND is shown in Figure 2.19.

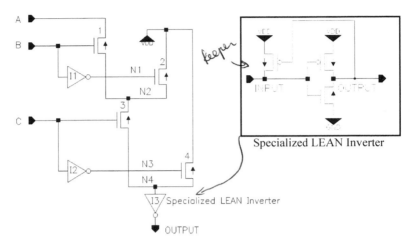

Specialized LEAN Inverter

FIGURE 2.19 LEAP Pass-Gate Logic of AND Function.

Function

Logical input B conditions pass NFET device 1, and its inverse conditions pass NFET device 2 to propagate either logic input A or V_{DD} to node N2. Logical input C conditions pass NFET device 3, and it's inverse preconditions pass NFET device 4. Together, they propagate the result of the first result, or again V_{DD}, to node N4, the input to the LEAN inverter/buffer. The output of the LEAN buffer produces the circuit's result. The specialized LEAN inverter is essentially a simple inverter with a half-latched "Keeper" PFET device which brings the inverter input the rest of the way up from V_{DD}-V_T to V_{DD}, as shown in the inset of Figure 2.19.

Characteristics

Strengths	Weaknesses
High speed operation	Non-Boolean Logic synthesis needed
Small, simple element library needed	Limited fan-out capability
High logical width and depth	No inverted signal availability
	Wider performance/process distribution

TABLE 2-12 LEAP Attributes

LEAP Logic addresses the issues associated with implementing pass gate logic in an automated design methodology. Traditional design methodologies attempting to execute functions of limited scope in pass gates realize minimal advantage over static logic. LEAP Logic maintains logic flexibility in order to support general logic function solutions, which do not necessarily resemble the standard Boolean solution reached with books of higher function. Indeed, building logic with 2-way pass gates is a very different paradigm. Flexibility comes from the simplicity of its library.

In essence, a function built in LEAP logic forms a tree, comprising a sequence of 2-way multiplexers. The path is similar to NFET pass gate logic (see "CMOS Pass Gate (PG) and Transmission Gate (TG) Logic" on page 68), but is differentiated by its attempt to implement whole functions via automated logic synthesis software, rather than independent logic books[13]. Logic is not generated differentially, in contrast to its CPL relative. Because the logic is single-ended, inverted signals are available only at the end of each tree, at the output of the inverter, or must be made available by inverting selected nodes.

---

13. A "book" is the name given to a simple logic function which has been built in a standard layout convention. The book is put into a virtual library, and used in multiple instances throughout a chip.

Non-Clocked Logic Styles

2.4.8 Double Pass-Transistor Logic (DPL)

Double Pass-Transistor Logic replaces the multiplicity of inverting buffers needed in CPL with redundant paths executed with complementary devices [2.21]. The Double Pass-Transistor Logic (DPL) implementation of a NAND is shown in Figure 2.20.

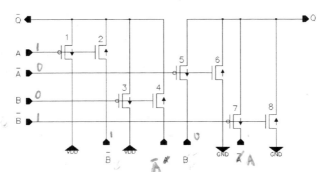

FIGURE 2.20 Double Pass-Transistor Logic implementation of AND function

Function

With inputs A and B low, devices 1 and 3 are both on, passing V_{DD} directly to the NAND output $\overline{Q}$. Devices 6 and 8 are also on, together driving AND output Q to ground.

Characteristics

Strengths	Weaknesses
High speed operation	Limited logic depth
Avoids buffer, redrive delay	Limited load capability
Avoids threshold-voltage drops within tree	Redundant device structure
Balance reduces data-dependent delay	

TABLE 2-13 DPTL Attributes

Double Pass-Transistor Logic, like CPL and DCVSL, is a dual-rail logic family which produces true and complement logic outputs from each stage. It enjoys high performance due to the low load capacitance it presents, but the load it drives must be limited. By requiring each input to connect to device sources as well as gates, more uniform delay and capacitive load is achieved on logical inputs. In other pass-gate arrangements, there can be substantial performance variation depending on the input data states. Since either NFETs or PFETs can be applied as needed[14], the circuit action is active pull-up or pull-down; neither source-follower action nor threshold voltage drop degrades overdrive in DPL circuitry. Buffers are therefore not essential on every stage, but have been found to be important after approximately 3 stages [2.22].

Unlike CPL and CMOS PG, more than one current path develops the output, reducing the resistance between the driver and the output capacitance. The multiple paths come from the "double transmission" property DPL exhibits. True and complement versions of both inputs are always available; true or complement versions of both signals, or supply rail levels are passed symmetrically through PFETs or NFETS to the output, gated in such a way that the two paths compatibly drive the required logical signal. Because no threshold voltage drop occurs, DPL can be operated at lower voltages without excessive noise immunity loss.

2.4.9 Pass-Gate Circuit Family Process Sensitivities

Process tolerance affects pass-gate-based performance by altering the RC delay of the gate's output load through the device and interconnect resistance. RC delay is encountered during evaluation of inputs and generation of outputs; since *propagated* rather than *regenerated* levels are coupled to output buffers during evaluation, evaluation RC delay is critical to pass gate performance. Delay variability is described in more detail in Chapter 4 and is summarized here.

Resistance

1. Device on-resistance, determined by its transconductance, is influenced by a number of mechanisms discussed in Chapter <n>1. First-order variation is associated with tolerance in threshold voltage, channel length, and gate-insulator thickness found across chip, wafer and lot.

2. Pass gate device on-resistance is made worse by body effect, and so pass-gate-based structures will be especially responsive to substrate sensitivity

3. Pass gates transition relatively slowly through the peak operating conditions for hot carrier wearout. Hot carrier effects need to be carefully assessed, as resulting higher thresholds aggravate body-effect performance loss.

---

14. The availability of both signal polarities allows selection of the device, NFET or PFET, which will provide the most current.

---

4. Changes in the device's I-V relationship will be multiplied many times. Most of the pass gate styles can handle significant depth; DCVSPG pulldown paths, for example, are sequentially-tied NFET devices.

5. For the same reason, logic path lengths must be restricted in styles which do not automatically re-buffer signals.

6. If the distance between circuits grows too large, the interconnect's resistive voltage drop causes the most-positive-down-level (MPDL) of the signal to be excessive at NFET logic tree device's source diffusion. Because DCVS-PG logic tree pass gates will have elevated thresholds due to body effect[15], it is critical that the devices be provided with logical low levels as close to ground voltage as possible.

7. PFET-to-NFET threshold voltage and channel length tracking is very important to styles such as Section 2.4.4, "Swing-Restored Pass Gate Logic (SRPL)" on page 76; shifts in latch hysteresis can profoundly impact delay.

Capacitance

1. Device overlap capacitance is a substantial portion of the total device capacitance. Because pass-gate circuit loads tend to be light, variation in overlap capacitance will change the load seen by the evaluate tree. In addition, high overlap capacitance exhibits Miller coupling, which delays the stabilization of the output.

2. Interconnect capacitance variation, caused by tolerance in wire width or thickness, can also alter the capacitance presented to the evaluate tree.

3. Interconnect width and height tolerance also alters lateral coupling. While unclocked pass-gates avoid the logic noise sensitivity of clocked pass gates (see "Clocked Pass-Gate Logic" on page 124), changes in coupling assert "delay noise" on the path, and can cause a set-up time into a latch to be missed.

Clearly, the vulnerabilities of prior styles are only a subset of pass-gate's sensitivities. With prudent margin and sound design technique however, pass-transistor-based logic can be made to be robust.

2.5 Summary

A survey of non-clocked static, DCVS, and pass-gate circuits illustrates some of the options available for implementing logic. These circuit topologies generally are reli-

---

15. See Section 1.2.9, "Body Effect" on page 21.

able and diagnosable, and resistant to logic noise. Further, these styles interface well with one another; a number of different styles can be intermixed.

The sensitivities of non-clocked circuit styles generally affect performance rather than functionality. The performance of static designs responds strongly to changes in FET device current caused by application conditions as well as process variation. DCVS styles, in addition, depend on close tracking of NFET to PFET device relative currents. Pass-gate-based designs are sensitive to those parameters plus the body effect imposed by elevated source voltages. The job at hand, then, is to assess the magnitude of total variability, so that appropriate margin can be allocated. This is explored in detail in Chapter 4.

Delay associated with the "overhead" that non-clocked circuits need to include can be large; these topologies have difficulty matching the performance of circuits which are optimized for monotonic operation. These circuits, as we will see in Chapter 3, are operated in only one direction, and have removed from their evaluate path many of those elements which increase delay, such as high fan out capacitance, latch hysteresis, and cross-over current. The overhead, although removed from the critical path delay, is larger than that found in static circuits. Determining design priorities, and then selecting appropriate circuit families, has never been more critical to a robust product.

REFERENCES

[2.1]   C. Mead, L. Conway, "Introduction to Vlsi Systems," 1979

[2.2]   I. Sutherland, etal., "Logical Effort: Designing for Speed on the Back of an Envelope," *Advanced Research in VLSI 1991: UC Santa Cruz*, pp. 1-16.

[2.3]   C-L Chen, etal., "Pulsed Static CMOS Circuit," *U.S. Patent # 5,495,188*, February 27, 1996

[2.4]   E. B. Eichelberger, etal., "A Logic Design Structure for LSI Testing," *Proceedings of 14th Design Automation Conference*, pp. 462-468, June 1987.

[2.5]   C. Mead, L. Conway, "Introduction to Vlsi Systems," 1979

[2.6]   L. G. Heller, etal., "Cascode Voltage Switch Logic: A Differential CMOS Logic Family," *Proceedings of 1984 IEEE International Solid-State Circuits Conference*, pp. 16-17.

[2.7]   R. K. Brayton, etal., "Decomposition and Factorization of Boolean Expressions," *Proc. IEEE ISCA*, Rome, Italy, May, 1982.

[2.8]   Leo C. M. G. Pfennings, etal., "Differential Split Level CMOS Logic for Subnanosecond Speeds," *IEEE Journal of Solid-State Circuits*, Vol. SC-20, No. 5, October, 1985, pp. 1050-1055.

[2.9]   S-L L. Lu, "Implementation of iterative arrays with CMOS differential logic," *IEEE Journal of Solid State Circuits*, Vol. 23, No. 4, August, 1991, pp. 1013-1017.

[2.10]   J-S Wang, etal., "CMOS Non-threshold Logic (NTL) and Cascode Nonthreshold Logic (CNTL) for High-Speed Applications," *IEEE Journal of Solid State Circuits*, Vol. 24, No. 3, June 1989, pp. 779-786.

[2.11]   M. Shoji, "FET scaling in domino CMOS gates," *IEEE Journal of Solid State Circuits*, Vol. 24, No. 3, June 1989, pp. 779-786.

[2.12]   R. Zimmermann, etal., "Low-Power Logic Styles: CMOS Versus Pass-Transistor Logic," *IEEE Journal of Solid State Circuits*, Vol. SC-20, No. 5, October 1985, pp. 1067-1071.

[2.13]   T. Dillinger, "VLSI Engineering," *Prentice-Hall Publishing Co.*, 1988, pp. 433-435.

[2.14]   Fang-shi Lai and Wei Hwang, "Design and Implementation of Differential Cascode Voltage Switch with Pass-Gate (DCVSPG) Logic for High-Performance Digital Systems," *IEEE Journal of Solid-State Circuits*, Vol.32, No. 4, April 1997, pp. 563-573.

[2.15]   K. Yano, etal., "A 3.8 ns CMOS 16 x 16 multiplier using complementary pass transistor logic," *IEEE Journal of Solid-State Circuits*, Vol.25, No. 2, April, 1990, pp. 388-395.

[2.16]   A. Parameswar, etal., "A High Speed, Low Power, Swing Restored Pass-Transistor Logic Based Multiply and Accumulate Circuit for Multimedia Applications," *Proceedings of 1994 IEEE Custom Integrated Circuits Conference*, pp. 278-281.

[2.17]   Kazuo Yano, etal, "Lean Integration, Achieving a Quantum Leap in Performance and Cost of Logic LSIs," *Proceedings of IEEE 1994 Custom Integrated Circuits Conference*, pp 603-606

[2.18]   Kazuo Yano, etal., "Top-Down Pass-Transistor Logic Design," *IEEE Journal of Solid-State Circuits*, Vol. 31, No. 6, June 1996, pp. 792-803.

[2.19]   M. Song, etal., "Design Methodology for High Speed and Low Power Digital Circuits with Energy Economized Pass-transistor Logic (EEPL)," *Proceedings of 22nd European Solid-State Circuits Conference*, Neuchatel, Switzerland, Sept. 1996, pp. 120-123.

[2.20]   W-H Paik, etal., "Push-pull Pass-transistor Logic Family for Low voltage and Low power," *Proceedings of 22nd European Solid-State Circuits Conference*, Neuchatel, Switzerland, Sept. 1996, pp. 116-119.

[2.21]   M. Suzuki, etal., "A 1.5 ns 32-b CMOS ALU in Double Pass-Transistor Logic," *IEEE Journal of Solid-State Circuits*, Vol.28, No. 11, November 1993, pp. 1145-1151.

[2.22]   V. Oklobdzija, etal., "Differential and Pass-Transistor CMOS Logic for High Performance Systems," *Proceedings of the 21st International Conference on Microelectronics, Vol2*, Nis, Yugoslavia, September 1997, pp. 803-810.

CHAPTER 3   *Clocked Logic Styles*

3.1 Introduction

In the preceding chapter, nonclocked circuit topologies were shown generally to be versatile, reliable, and relatively low in power consumption. Clocked logic, on the other hand, is recognized for its performance advantages, which may be attributed to the following:

1. In Static CMOS, logic must be built redundantly; circuit operations must be realized in both NFET and PFET device structures to accommodate both up and down logic transitions. This reduces performance by adding **gate** fan-out load and interconnect RC. Higher device counts lead to longer interconnects, higher power consumption and bigger die[1].

2. In static CMOS, even when the redundant structure is off, the added **diffusion and overlap** capacitive loads increase power and delay.

3. In Static CMOS, PFET devices must drive the same loads as NFET devices, at half the transconductance. This drives PFET devices to generally be 2X the width of NFET devices for balanced transitions. The impact is of particular concern structures such as PFET devices 5 and 6 in Figure 2.2a.

---

1. See Figure 2.1, ""Design Explosion" vs. "Design Convergence"," on page 53

Because *clocked* logic states are developed in only one direction, a single "device polarity" (usually NFET) is used predominantly in the evaluate path. Since the device size is optimized for that transition, capacitance can be dramatically reduced. There are less devices to drive, and those devices tend to be smaller.

4. The I_{DS} of logic evaluation devices in clocked circuitry is usually devoted 100% to switching logic states, rather than also needing to sink static "crowbar" currents [3.1] (See "Static CMOS Structures" on page 54.).

5. The switchpoint of a static circuit stage typically occurs at voltage $V_{DD}/2$. Clocked dynamic circuits, however, switch with substantially less voltage swing. And as previously described, when it does switch, less capacitance is charged.

The cost of clocked logic performance is a restraint in some product applications:

1. An inherent race exists between clock and data. In static logic, the only existing race was the potential for late logic arrival at the target latch after the close of the data setup period. In clocked logic, the data chases the clock at the circuit stage as well as at the latch, via multiple scenarios.

2. Logic clocking can substantially increase power consumption, and may account for 20% or more of total chip power.

3. The additional area needed to buffer or condition dynamic summand nodes increases total device count and critical defect-sensitive area, which decreases product yield.

4. Dynamic logic is intrinsically unstable. The logic summand node is vulnerable during evaluate to a number of process-induced and design-induced failure mechanisms. Leakage[2], noise[3], missed timings[4], and application conditions[5] can reduce the reliability of a design if not carefully assessed.

With rigorous design checking and verification, the performance improvement provided by clocked structures[6] can be equivalent to the gain realized by a full CMOS technology migration. While it is impractical to include every CMOS clocked logic

2. See Section 1.3, Charge Loss Mechanisms on page 25.

3. See Section 4.5, "Noise" on page 157

4. See Chapter 8.

5. See Section 4.4, "Application Induced Variation" on page 153

6. Adiabatic and charge-recovery-based clocked logic styles are not covered in this text. While their power consumption attributes deserve examination, few of the proposed structures yet produce interesting high speed logic performance.

structure appearing in the literature, a survey of significant structures is covered in Section 3.2 through Section 3.6.

3.2 Single-Rail Domino Logic Styles

Domino CMOS

Multiple Output Domino (MODL)

Compound Domino

Noise-Tolerant Precharge Domino (NTP)

Clock-Delayed Domino Logic

Self-Resetting Domino Logic

Clocked domino is the most common form of dynamic logic, enjoying a 20-50% performance advantage over static circuitry. Let's first examine a common domino structure to understand the style and its strengths.

3.2.1 Domino CMOS

The single-ended dynamic domino structure is common in recent high speed logic designs. Domino CMOS, proposed by Krambeck, etal, in 1982 [3.2], uses an inverter to resolve a critical race: logical inputs to a given domino circuit from a prior stage and cycle may corrupt the next cycle's precharge before new valid inputs can be delivered. Domino CMOS provides a straightforward solution. The single-ended dynamic domino realization of a two-way AND function is shown in Figure 3.1a.

Function

Dynamic dominos are composed of precharge, evaluate, and buffer functional blocks. Referring to Figure 3.1a, during precharge clock PC is low. PFET device 1 charges node N1 to V_{DD}, driving output Q to ground and turning on "keeper" PFET device 2. "Foot switch" NFET device 5 is off, interrupting the path to ground during precharge of the evaluate block. The evaluate block in Figure 3.1a is represented by devices 3 and 4. When PC transitions high, the circuit switches from precharge to evaluate mode. With logic inputs A and B high, node N1 is coupled to ground, switching the output of inverter buffer I1 high and turning off keeper PFET device 2. Before examining the waveforms of a domino operation, lets first introduce another important domino circuit concept.

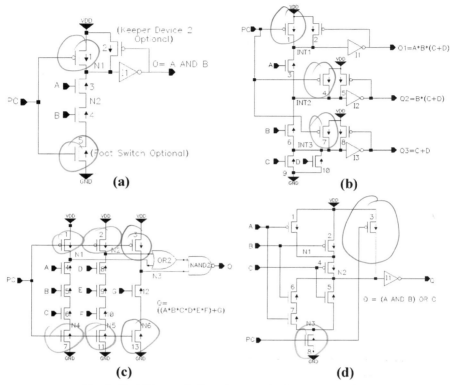

FIGURE 3.1 Single-Rail Dynamic Domino topologies: (a) Domino CMOS; (b) Multiple-Output Domino (MODL); (c) Compound Domino; (d) Noise-Tolerant Precharge (NTP).

Characteristics

Strengths	Weaknesses
Superior logic density	Logically incomplete circuit family
High performance	High clock power
Very low noise generation	Low noise immunity
Non-glitching output	High switch factor
	Limited fail diagnosability

TABLE 3-1 Domino CMOS attributes

Single ended dominos can contain a substantial amount of logic "width" in the NOR direction, and "depth" in the NAND direction. This allows significant "logic gain" along a path of dominos. Further advantages are found in sequences of dominos: if logic inputs A and B are outputs of other domino circuits using the same clock, then during precharge, they are also low and the footswitch, NFET device 5, which increases delay by adding resistance to ground, may be omitted.

Single-ended domino is an incomplete logic family; the structure is incapable of inverting logic. If the inverted version of the logic is needed along the path, a separate path originating from the last latch must be built. This separate path propagates inverted logic from the complement output of the last latch. The complement path runs in parallel with the true path to make the inverted output available where needed. As we will examine in Section 3.4.1 on page 111, dual rail, or *differential*[7] domino propagates true and complement logic together to execute complete logic functions.

Although quite fast, dominos generate little noise, as switching during evaluate is accomplished by small inverter/buffers, driven by the discharge of lightly loaded summand nodes which switch in one direction only. The launch of the initial stage in each logic path using a given clock edge is synchronized, and so generates the highest currents in the supply and signal interconnects. After that, the switching activity quickly becomes less synchronized.

Once the summand node is discharged past the switchpoint of the output inverter/buffer[8], there is no opportunity to revert the circuit back to its precharged state until the arrival of the refresh edge of the clock. Although unintentional discharge can be a problem, this response mechanism eliminates the "glitching[9]" seen in static paths. The keeper, or holding device, PFET device 2, is an optional device which provides replacement charge for leakage loss. The keeper, with the output inverter/buffer, forms a "half-latch," adding delay to the circuit by introducing hysteresis. The keeper can not be sized large enough to respond to noise events and still maintain performance advantages over nonclocked static implementations. If the circuit is precharged *every* clock cycle, and if the clock never quiesces the domino stage in its evaluate mode, then the keeper is sometimes omitted. As we will see, the keeper can

---

7. See Footnote 4 on page 59

8. The summand node may lose its precharge to logic evaluation, to process leakage paths (see page 26), or to design deficiencies (see page 96).

9. "Glitching" refers to the generation of intermediate transient output states by non-clocked circuits during the resolution of intersecting logic paths of varying duration. See "Static CMOS Structures" on page 54.

---

actually improve performance by preventing capacitive up-coupling of the evaluate node.

> **Rule of thumb:** Keeper PFET device width/length ratios are typically 1-20% of the composite width/length ratio of the domino's evaluate tree.

The designer has the option to "push aside" the output inverter I1 in Figure 3.1a, taking the domino's output from node N1 to improve performance. With severely limited output drive capability, this node is vulnerable to a number of problems. This practice is hazardous, and the author has often seen functional fails in hardware caused by this structure.

Design Sensitivities

With this understanding of domino operation, lets now examine this style's sensitivities. The integrity of this clocked structure is directly associated with the soundness of the precharge on the evaluate summand node capacitance. As discussed in Section 1.3, "Charge Loss Mechanisms" on page 25, *process and technology* limitations create mechanisms which erode the magnitude of charge on the summand node capacitor. In addition, certain *design practices* can also reduce precharged node voltage. A few examples are offered:

- *Capacitive charge divider* can reduce precharged node voltages. In the domino circuit shown in Figure 3.1a, assume N2, an intermediate node of an evaluate tree, was left at ground potential due to the prior evaluate. If logic input A goes high but logic input B remains low during evaluate mode, then even though no path to ground is established, the voltage on node N1 will dip as its precharge redistributes between the capacitance of node N1 and N2. The capacitance on node N2 is composed of the capacitance across node N2's diffusion junction to substrate, and overlap of the diffusion by the gates of devices 3 and 4. In larger domino evaluate trees, additional PFET precharge devices, also driven by the same PC clock, are inserted to precharge these intermediate nodes, avoiding this loss of precharge voltage.

- *Ground bounce* becomes a serious concern in dynamic circuits. The valuable precharge on the summand node of the evaluate tree must be preserved in a robust design. If transients on the ground plane can temporarily pull the source of evaluate NFET device 5 in Figure 3.1a below ground by a voltage approaching the NFET device threshold voltage, device 5 can momentarily turn on, eroding precharge. The synchronous nature of clocked logic easily can produce these transients, and experience has shown this to be a common fail mode. This form of noise is discussed in more detail in Section 4.5.5, "Simultaneous Switching" on page 167.

- *Interconnect coupling* allows signal transitions occurring on neighboring interconnect lines to capacitively couple high the interconnect of gates in dynamic circuits.
  If the coupled voltage approaches in magnitude an NFET device threshold voltage, precharge can be lost, causing false outputs. Coupling noise is seen often, and
  is difficult to diagnose. Interconnect delay noise and logic noise are addressed in
  more detail in Section 4.5.3, "Logic Noise" on page 160.

- *Bootstrapped diode clipping* causes loss of precharge in an unusual manner. If the
  uppermost logic input in the logic tree (gate of NFET device 3 in Figure 3.1a) couples high due to noise, Miller capacitance between the gate and the drain of NFET
  device 3 can "bootstrap" the precharged summand node high. If the overlap capacitance is sufficiently high and the summand node capacitance is sufficiently low,
  the summand node may rise in voltage above V_{DD} sufficiently to forward bias the
  drain diode of the precharge device. When the Miller capacitance is eventually
  satisfied, the summand node returns to a lower voltage with less noise margin,
  reflecting the charge loss through the diode. The domino's signal margin remains
  vulnerable to subsequent noise until the next precharge. The presence of a keeper
  PFET can prevent this action by rigidly holding the node to voltage V_{DD}.

- *Minority charge injection* of electrons into the p-type substrate is caused by
  "undershoot" of strongly driven signals. When the down-going slew rate of a signal tied to a node exceeds the rate at which that node can be capacitively discharged, the node is pushed negative with respect to the substrate. This event
  increases the amount of minority charge liberated into the substrate and swept up
  by the positively-precharged diffusion diode. Recombination of this charge occurs
  as it is swept up, consuming precharge and reducing charge available to develop
  the output signal.

- *Operating temperature* strongly influences the charge lost to subthreshold leakage
  current. Leakage increases exponentially with temperature, threatening the integrity of dynamic circuits with wider logical widths. Designs must anticipate the
  temperature extremes of the application conditions.

Because domino stages are not actively held, diagnosis of domino path fails can be
frustrating. Static evaluation can be introduced into dynamic circuits to support more
conventional directed diagnostic techniques. As shown in Figure 3.2, PFET device 6,
sized quite small, is turned on during test or stress to operate the circuit statically.
This practice introduces relative minor additional evaluate capacitance, but potentially substantial additional test /stress power consumption and device wearout.

Generally, domino circuit styles are not "naturally" low power circuits. Clock infrastructure is required to assert precharge and evaluate timings on the circuits. In addition, the switch factor of a domino is high. From cycle to cycle, even if the output

---

Clocked Logic Styles   **97**

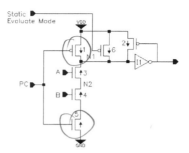

FIGURE 3.2 Static evaluate configuration for dynamic domino

state of a given domino circuit does not, say, go high, it still must precharge and then re-evaluate the same inputs. Nonetheless, logic evaluate structures are not replicated in PFETS. In contrast, static circuits will not switch over a cycle if their inputs don't change, avoiding additional AC power consumption. The logic structure appearing in NFETs must be recreated in PFETs, however.

> **Rule of thumb:** Generally, dynamic circuits consume 3X - 5X the power of static circuits.

3.2.2 Multiple Output Domino Logic (MODL)

Multiple Output Domino Logic (MODL) exploits the availability of intermediate evaluate tree outputs [3.3]. By conditioning intermediates as outputs, savings in power, area, and performance are realized by avoiding reproduction of subsets of the given logic tree. The MODL topology realizing the 2-way OR and the, 2-way AND and 2-WAY OR calculation is illustrated in Figure 3.1b.

Function

Assume no foot-switching is needed. Once clock PC rises, the precharged summand nodes INT1, INT2, and INT3 shown in Figure 3.1b are precharged and floating at voltage V_{DD}; the MODL circuit is now said to be in evaluate mode. If logic inputs A and D remain low while inputs B and C transition high, node INT1 remains high, while nodes INT2 and INT3 are discharged through NFET evaluate devices 6 and 9, respectively, to ground. The output of inverters I1 and I2 transition to a high value until precharge clock PC again transitions low; inverter I3's output remains in a low state. Figure 3.3 illustrates the MODL circuit's described operation sequence. Note the noise induced on the internal nodes, which must be managed, when higher tree inputs transition high.

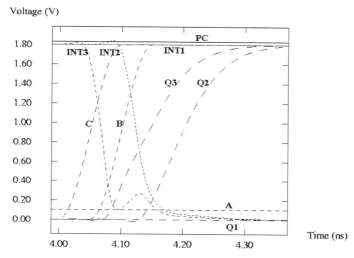

FIGURE 3.3 Operation of MODL circuit in Figure 3.1b.

Characteristics

Strengths	Weaknesses
Active power reduction	Logically incomplete circuit family
Low noise generation	Larger evaluate capacitance
Area, device count reduction	Additional noise concerns
	High switch factor

TABLE 3-2 MODL attributes.

Multiple output domino provides area and delay reductions by accomplishing more than one logic operation in the same evaluate tree, evaluating and outputting the voltages of intermediate evaluate tree nodes. The logic must be ordered in its tree position such that the intermediate results are subsets of the greater function accomplished by the entire circuit. In Figure 3.1b, two stage delays and 16 transistors are required to develop the relatively simple outputs Q1, Q2, and Q3. Five devices have been eliminated from independent implementations. In many cases, up to half the number of devices can be saved. It has been demonstrated that delay goes linearly with tree fan-in [3.4].

A trade-off is found in the device width increase needed in the entire evaluate tree to sink the currents of multiple precharged nodes. These increased devices present incrementally higher fan out capacitance to drive, and more evaluate tree capacitance to discharge. In addition, as the intermediate nodes introduce more loading, clocked PFET precharge devices on these nodes become mandatory. This added overhead has been shown to be small, enabling MODL to be effective in reducing delay and area, specifically in adder designs [3.4].

A second performance trade-off is associated with the intermediate precharging. In dominos associated with critical paths, the designer may sometimes choose to *predischarge* rather than *precharge* intermediate to enhance pull down performance, in scenarios where the resulting charge sharing has already been shown to be tolerated. This aggressive design practice is prevented in MODL.

Finally, MODL structures can create unintended current paths through the subset logic causing intermediate node discharge and false outputs. Attention must be paid to the timing and order of the logical inputs to ensure that the outputs are robust chronologically as well as logically.

3.2.3 Compound Domino Logic

The compound domino circuit shown in Figure 3.1c introduces two important extensions of the domino circuit style [3.5].

1. Due to the capacitive precharge divider concern described on page 96, conventional domino's evaluate tree height is limited due to the number of intermediate nodes which may potentially consume summand node precharge. The designer is discouraged from adding intermediate clocked precharge devices due to their performance impact. But by splitting the inputs into multiple, independently precharged sections, charge sharing can be reduced, and noise immunity loss is mitigated.

2. Common domino logic requires inversion of the precharged summand node so that inputs to subsequent dominos are low during precharge. Compound domino structures demonstrate this inverter can be replaced by other inverting static CMOS logic circuits, in order to achieve more complex logic functions.

Function

Referring to Figure 3.1c, Clock input PC precharges independent summand nodes N1, N2, and N3 when in the low state by turn on PFET devices 1-3 and turning off NFET devices 7, 11, and 13. When PC transitions to the high state, NFET devices 7, 11 and 13 turn on and allow the three independent logic trees formed by NFET devices 4-6, 8-10, and 12, respectively, to evaluate logic inputs A-G. Summand node outputs N1, N2, and N3 are then passed to static logic gates OR2 and NAND2 to logically recombine the separate summands back into a single domino output.

Characteristics

Strengths	Weaknesses
Superior logic density	Logically incomplete
Low noise generation	Additional evaluate delay
Reduced charge sharing	Low noise immunity
Limited static logic circuitry usage	High switch factor

TABLE 3-3 Compound Domino attributes.

Compound domino is a handy idea in the domino "toolbox," as it is compatible with common domino logic. Besides addressing the charge sharing concern discussed on page 96, this structure can also enhance performance in a surprising manner. Typically, the critical delay defining the minimum cycle time of a processor is the evaluation of data. In large, logically wide and deep domino paths, however, it is possible for the precharge delay to become the predominant consideration in minimum cycle time, as the number of source and drain diffusions increases. Compound domino is one means of reducing the number of diffusions in a tree needing precharge, hence cutting precharge delay.

Depending on logic function required, it is also possible to bring an already inverted signal, if already available, directly into the static output logic. Compound domino already has increased the complexity of the output logic, so additional inputs only increase delay incrementally. The delay of the static output logic must be monitored closely however, so as to not compromise the gains realized by domino action. The delay associated with the more complex output logic should be expected to exceed that of the inverter it replaces.

3.2.4 Noise Tolerant Precharge Logic (NTP)

The domino circuit's vulnerability to noise is associated with precisely that feature which makes it so fast. As outlined in section 3.2.1, noise coupled into the domino during the evaluate portion of its cycle can cause development of false output states, as the domino's precharged summand node is left to float. Noise Tolerant Precharge (NTP) logic provides higher noise immunity by avoiding floating evaluate nodes condition during evaluate [3.6] [3.7]. The NTP structure accomplishing the ((A and B) or C) operation is shown in Figure 3.1d.

Function

Referring to Figure 3.1d, Clock input PC is low during precharge, charging summand node N2 to the high state via PFET device 3, with NFET device 8 off. When PC transitions to the high state, NFET devices 8 turns on, allowing logical inputs A-C to develop the voltage of summand node N2 via NFET devices 5-7. Summand node output N2 is then inverted by inverter gate I1 and passed as the output of the circuit. PFET devices 1, 2, and 4 provide active noise immunity by providing CMOS-type complementary logic pull-up.

Characteristics

Strengths	Weaknesses
Elimination of floating node concerns	Larger fan-out load
High noise immunity	Monotonic logic propagation
High failure diagnosability	Higher device count, wiring complexity
Elimination of half-latch	

TABLE 3-4 NTP attributes.

Noise Tolerant Precharge logic never allows a precharged node to float, resembling in many ways the static equivalent of that function. Complementary PFET logic is always on when the NFET evaluate path to ground is off. In the preferred embodiment, PFET devices 1, 2, and 4 in Figure 3.1d are not large enough to accomplish the precharge function, but are sufficient to provide a high degree of noise immunity. This level of noise immunity is not usually seen with common dominos: hysteresis from keeper devices which are large enough to affect noise margin will substantially degrade performance. The now-unnecessary keeper half-latch is so eliminated, avoiding the brief contention between the NFET pull-down/evaluate path and the pull-up current from the keeper.

The trade-off for this added noise immunity and performance is, again, similar to the costs of the static circuit that NTP resembles. Increased area and fan-out capacitance will increase power consumption. Output inverter/buffers need to be sized to accommodate the large fanout load in order to preserve performance.

3.2.5 Clock-Delayed Domino (CD Domino)

An important single-ended, self-timed logic style, Clock-delayed (CD) Domino[10], eliminates the fundamental monotonic signal requirement imposed on standard

---

10. Clock-Delayed Domino is also referred to as *Delayed Reset Domino.*

dynamic domino, by propagating a clock network in parallel to the logic, providing a dedicated clock to each logic stage [3.8]. Clock-delayed domino logic was heavily used in the highest performance microprocessor reported to date [3.9]. A simple CD domino stage is shown in Figure 3.4a.

Function

The example CD Domino stage comprises a standard 2-way OR executed in single-ended domino with transistors 1-5 and inverter I1, and a clock delay element formed by inverters I2 and I3, and transmission gate transistors 6 and 7. When the precharge/evaluate clock transitions to the high state, device 5 turns on, and the domino circuit evaluates the OR of logical inputs A and B. After inverter I1 has fully developed output Q, the delay circuit produces a delayed clock output which is provided to subsequent stages.

Characteristics

Strengths	Weaknesses
Capability of inverting functions	Extreme process sensitivity
Reduced chip clock overhead	Timing complexity
Higher density, including delay elements	Additional clock path propagated with logic

TABLE 3-5 CD Domino attributes.

Because of the presence of dedicated clocks, CD domino provides both inverting and non-inverting functions, without the need for differential pairs or cross-coupled latches. Inverter I1 in Figure 3.4a is not essential, because the clock will always arrive *after* the interval when evaluate should have been completed. Two timing schemes are described in the literature:

> In the first approach, a single delayed clock is provided to *all* domino gates occupying the same sequential position in their respective logic paths. The clock delay is large enough to allow evaluation of the slowest of the gates. Power consumption, as with standard domino, remains highly synchronized, and designs must accommodate substantial simultaneous-switch noise.[11]

> In the second approach, *each* domino gate has generated in parallel a clock, with delay slightly longer than the slowest logic path through that stage[12]. Extra

---

11. "Wave Domino" executes paths of logic in this manner [3.10].

12. Clock delay adjustment is accomplished by varying the size of transmission gate devices 6 and 7, or by adding additional transmission gate devices in series with devices 6 and 7.

---

devices are required, but performance is improved. In addition, lower peak currents and less simultaneous switching is observed, as clock action is less synchronous.

Clock delay specific to the path ensures that all inputs have stabilized and that the
domino has had time to complete it's evaluation. Design margin must be added to the
clock delay to guarantee function across process, circuit, and application variations[13].
CD domino's performance and scalability will cause the style to become increasingly
popular.

3.2.6 Self-Resetting Domino (SRCMOS)

It's interesting to read the patent first describing Self-Resetting CMOS [3.11]. The
ability to achieve faster high-to-low transitions in precharged PLA domino circuits is
highlighted, by gating pull-up current without a clock. Indeed, it does achieve that
function; but more importantly, SRCMOS has introduced the use of high-speed
dynamic logic in whole semi-synchronous or asynchronous designs. Unlike the previous dynamic logic concepts described above, SRCMOS structures generically detect
their operating phase and supply their own appropriate clocking. A self-resetting realization of a 3 way-OR is shown in Figure 3.4b.

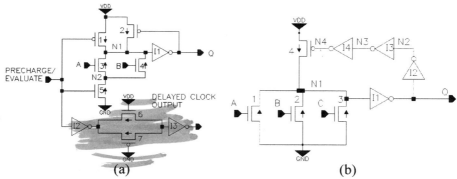

(a)   (b)

FIGURE 3.4 Self-Timed Single-ended Domino Structures; Clock-Delayed Domino
(a); SRCMOS (b).

---

13. Performance variability is examined in detail in Chapter 4.

Function

Referring to Figure 3.4b, assume logic inputs "A", "B", and "C" are low. NFET logic devices 1, 2, and 3 are all off. Node N1, N2 and N4 remain high, and output Q as well as N3 stay low. Assume a high logical input pulse is coupled to "A"; node N1 falls, and output Q goes high, as shown in Figure 3.5. After 3 sequential inverter delays, node N4 goes low to precharge node N1 back up to V_{DD}. By the time this precharge is re-initiated, logical input "A" has already returned low. The circuit returns to its resting state; node N1 is precharged, output Q is low, and PFET precharge device 4 is off.

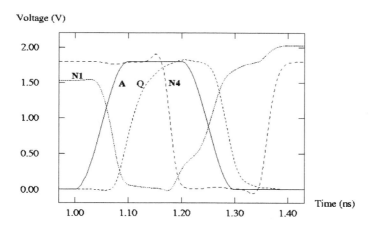

FIGURE 3.5 Wavefront passing through SRCMOS element shown in Figure 3.4a.

Characteristics

Strengths	Weaknesses
Very high performance	Difficult to time
Reduced clock overhead	Low noise immunity
Evaluate/clock race avoided	High process sensitivity
	Difficult fail diagnosability
	Additional device count

TABLE 3-6 SRCMOS attributes.

SRCMOS acquaints the reader with the notion of propagating logic as a *wave* or *bubble*, rather than as a *level*. Unlike other styles, the output values are available for only a limited duration; the restore wavefront chases the logic evaluation wavefront-through the logic path. The discharge of the dynamic summand node automatically initiates a delayed restore after enough time has been allowed for the logic wavefront to move through the circuit. The input signal to a SRCMOS stage is a pulse rather than a level, whose duration must be less than the reset delay, in this example the propagation delay through inverters I1-I4.

SRCMOS designs are extremely fast and dense, but also quite difficult to time. Inputs to a given SRCMOS stage must arrive within a specific interval, made even more difficult by the high process sensitivity of the style. Diagnosing fails is complicated by the transience of the signals. An alternative structure resolves this by gating the reset inverter chain during test to support static evaluation and diagnostics [3.12]. Finally, noise can be propagated along with data, and so must be considered during layout. Self-resetting circuitry is intrinsic to the asynchronous systems discussed in more detail in Section 7.7, "Asynchronous Techniques" on page 281.

Self-resetting structures save power in two ways. When data present at evaluate does not require dynamic node discharge, the precharge device is not active, as it would be in clocked domino. In addition, the clock infrastructure is now limited to the latches which launch and receive the path, eliminating clock distribution wire load. Additional power is spent, however, in the three added inverters and in scenarios where the input pulse is longer than 3 inverter delays, causing DC current. A somewhat larger self-resetting style saves power by gating the pull-up current earlier, with the "OR" of the logical inputs to the domino [3.13]. The structure, shown in Figure 3.6, also decreases alpha and cosmic ray error rates[14].

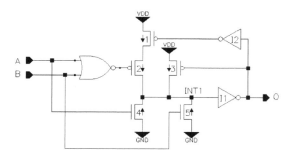

FIGURE 3.6 Self-restored domino with SER and power features.

---

14. See Section 1.3.4, "Alpha Particle and Cosmic Ray Interactions" on page 30.

3.3 Alternating-Polarity Domino Approaches

NORA Domino

Zipper Domino

Two similar single-ended structures eliminate domino's output inverter/buffer delay by alternating use of NFET and PFET devices in the evaluate tree. The performance improvement associated with uninverted logic propagation must be tempered by the lower transconductance associated with PFET evaluate devices found on half of the stages. A NORA path of arbitrary functions is shown in Figure 3.7a,.

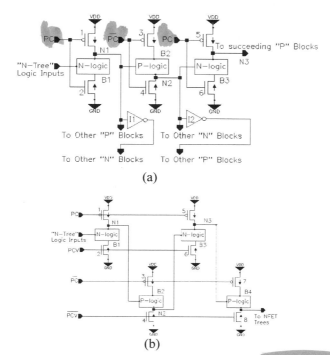

FIGURE 3.7 Alternating polarity Domino Structures: (a) NORA Dynamic Logic; (b) Zipper Domino.

3.3.1 NORA

Common Domino CMOS, although quite powerful, carries design liabilities. Inverters are required by the circuit to address polarity and race problems, but add path delay without developing new information. Domino CMOS also is logically incomplete, and so is much more efficient only for certain logic configurations. The NORA (NO RACE) technique allows the removal of the inverter without recreating the internal race [3.14]. NORA is often mistaken for Zipper CMOS [3.15], but the two have distinct differences. A representative NORA structure is shown in Figure 3.7a.

Function

Assume a logic path comprising 3 sequential NORA stages, shown in Figure 3.7a. During the cycle's precharge interval, Clock PC is at ground, and $\overline{PC}$ is at voltage V_{DD}, preconditioning nodes N1 and N3 high, and node N2 low. Foot-switch devices 2, 3, and 6 are off. When PC goes high, and $\overline{PC}$ goes low, the NORA path enters the evaluate period. Assume N-tree logical inputs transition high. Node N1 is hence pulled down through the N-logic tree B1, passing a low voltage input to P-logic tree B2. With input(s) N1 low, P-logic tree B2 pulls node N2 high. Finally, with input(s) N2 high, N3 is pulled to ground through N-logic tree B3. A series of NORA clock cycles is shown in Figure 3.8 below. Note that although logic trees B1, B2, and B3 for

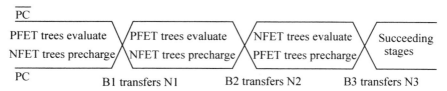

FIGURE 3.8 Operation of NORA circuit in Figure 3.7a.

simplicity have only one logical input in Figure 3.7a, these evaluate blocks are most efficient with multiple inputs.

Characteristics

Strengths	Weaknesses
Reduced intrinsic delay stages	PFET logic propagation delay
Lower fanout, parasitic capacitance	Additional clocking complexity
Smaller area	Low fan-out capability
	Restrictions on logic inputs

TABLE 3-7 NORA attributes.

NORA structures have the flexibility to provide both polarities of output when needed, and exceeds Domino CMOS in flexibility. In cycle-limiting paths, alternating between PFET and NFET evaluate-block circuits avoids the need for inversion: when intersecting logic vectors need inverted summands, intermediate outputs such as nodes N1-N3 can be coupled to small inverters; sequential NORA circuits of the same device type which are separated by inverters revert to conventional Domino CMOS. As with Domino CMOS, NORA circuit path length can be varied to suit the function. In addition, NORA logical outputs can be cleanly "handed off" to successive half-cycle paths containing NORA circuits clocked in opposite polarity, enabling uninterrupted evaluation.

The sensitivities of NORA logic are associated with the opportunities for precharge erosion, and are common to all dynamic logic families. The reader is directed to Section 1.3, "Charge Loss Mechanisms" on page 25, for a treatment of dynamic logic vulnerabilities. In summary, NORA structures must observe the following:

1. The physical distance and capacitive load must be minimized.
2. An additional clock, $\overline{PC}$ must be piped throughout the logic.
3. Noise margin becomes critical, as none of the nodes are buffered.
4. Unless inverters such as I1 and I2 in Figure 3.7a are added, N-tree outputs can be provided directly only to P-tree inputs, and P-tree outputs can be provided directly only to N-tree inputs, making logic synthesis somewhat more cumbersome.
5. Logic delay has a dependence on PFET as well as NFET transconductance.
6. Due to 5) above, the number of parametrics with first order critical path delay influence is approximately 2X that of Domino CMOS.

Once again, the merits of a potentially powerful circuit topology are strongly dependent on the characteristics of the process technology to be used.

3.3.2 Zipper Domino

In NORA, noise and charge loss are first order concerns. NORA cuts the precharged summand node capacitance by more than half, which improves precharge and evaluate performance, but challenges noise immunity and circuit robustness. Zipper Domino, shown in Figure 3.7b, advances the alternating n/p cmos concept by resolving the charge-sharing and noise margin concerns [3.15].

Function

Referring to Figure 3.7b, assume the zipper domino path is first in precharge mode. Clock inputs PC and PCV are at ground potential, and inputs $\overline{PC}$ and $\overline{PCV}$ are at voltage V_{DD}. PFET devices 1 and 5, turned on by PC, precharge summand nodes N1 and N3. NFET devices 4 and 8, turned on by clock $\overline{PCV}$, predischarge summand nodes N2 and Output to ground. Footswitch devices 2, 3, 6, and 7 are off, preserving the preset state. When the path switches to evaluate mode, Clock input PC rises to V_{DD}, clock input PCV rises to V_{DD}-V_T, clock input $\overline{PC}$ goes to ground voltage, and clock input $\overline{PCV}$ falls to voltage V_{TN}. Logic trees B1-B4 then evaluate. "N-Tree" compatible monotonic logic inputs, which are low during precharge, are provided to evaluate tree B1. The uninverted output of n-logic block B1, on node N1, is then only appropriate for p-logic trees, such as block B2 as shown in Figure 3.7b.

Characteristics

With the inverting buffers (inverters I1 and I2 in Figure 3.7a) removed from the NORA structure, the only opportunity to replace charge loss remains with the precharge device. By separating the clocks coupled to the precharge device and the foot switch, the bias on the precharge devices can be elevated, and subthreshold current will supplant leakage current much as the keeper had. The approach has none of the hysteresis the half-latch caused, but then also does not shut completely off when the summand node is discharged by the evaluate tree. This DC current in, say, 50% of the stages can consume substantial power. In addition, wide variation in the device current should be expected, due to normal tolerance[15], and the high gain associated with operating a device near weak inversion. Although the style introduces important concepts, Zipper CMOS never saw widespread usage in the industry.

---

15. See Section 1.2, Front-End-Of-Line Variability Considerations on page 6.

3.4 Dual-Rail Domino Structures

Differential Domino
Cross-Coupled Differential Domino
Modified Dual-Rail Domino
Dynamic Differential Split Level Logic
Pseudo-Clocked Domino Logic

In contrast to the previous structures, Dual Rail Domino structures are true, clocked *differential* circuits which directly generate *complementary* outputs. Dual-Rail Domino is the clocked antecedent of static DCVS logic: it implicitly inverts logical outputs, creating a complete logic family *using pairs of dominos* [3.16]. In addition, inputs can be evaluated without pass gate or transmission gate circuitry, which can compromise performance or timing accuracy[16]. Four Dual Rail Domino styles are shown in Figure 3.9.

3.4.1 Differential Domino

Differential domino is the original clocked embodiment of DCVS disclosed by Heller, etal [3.16]; the differential domino realization of a 2-Way AND circuit is shown in Figure 3.9a.

Function

Referring to Figure 3.9a, internal nodes N1 and N2 are preconditioned to voltage V_{DD} by PFET devices 2 and 3. When clock PC transitions high as shown in Figure 3.10, PFET devices 2 and 3 are cut off, and the footswitch, NFET device 9, turns on. With logic inputs A and B high, node N1 is discharged to ground, and node N2 remains high. Keeper PFET device 11 turns off, and PFET device I2 remains on. Output Q transitions high and $\overline{Q}$ remains low.

Characteristics

Dual Rail Domino's increased evaluate tree efficiency, stemming from device sharing, is apparent in Figure 3.9a. Sharing evaluate nodes improves power, area, and per-

---

16. Pass gates suffer from source-follower delay (See "Body Effect" on page 21.), and are vulnerable to threshold voltage variation associated with Short Channel Effect. (See "Device Threshold Voltage Variation" on page 13.)

Clocked Logic Styles

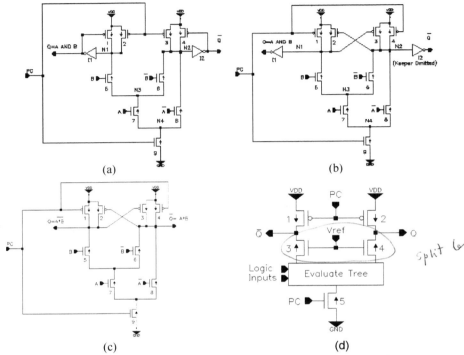

(a)   (b)

(c)   (d)

FIGURE 3.9 Dual-Rail Domino Structures: (a) Differential Domino; (b) Cross-Coupled Differential Domino; (c) Modified Dual Rail Domino; (d) Dynamic Differential Split-Level Logic.

formance. Alas, the structure still suffers from delay associated with overcoming half-latch hysteresis. A minor modification discussed below provides an opportunity to avoid that delay, arising from the circuit's differential nature. The following alternative still actively preserves node precharge.

3.4.2 Cross-Coupled Domino

In Figure 3.9b, a dual rail structure similar to differential domino achieves the same 2-Way AND logical function, but provides current replacement and noise margin by driving feedback from the complement side's summand node, rather than off the same side's inverted summand node. In this way, the delay required to overcome the switchpoint of the keeper halflatch is avoided, without sacrificing supplemental charge replacement and noise immunity during evaluate mode. This savings must be

Voltage (V)

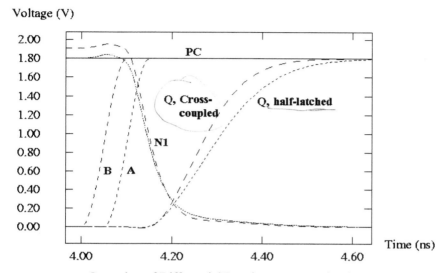

FIGURE 3.10 Operation of Differential Domino structures in Figure 3.9a and b.

weighed against the added delay associated with the gate capacitance of devices 2 and 3. Figure 3.10 demonstrates that the performance improvement associated with eliminating the half-latch exceeds the performance penalty associated with the additional cross-coupled device capacitance.

Characteristics

Strengths	Weaknesses
Logically Complete Family	Higher device count, area penalty
Enhanced noise immunity	Higher clock load
High performance	Higher power consumption/switch factor
Opportunity for self-timing	

TABLE 3-8 Differential Domino attributes.

The use of differential logic in dynamic structures is particularly appealing. By constructing dynamic logic evaluate trees in true/complement pairs, NFET evaluate devices may be shared, saving area in settings where both states were needed; in Figure 3.9a, NFET devices 7 and 9 are used for both trees. The use of selected precharge/charge-maintenance schemes, such as the cross-coupled PFET load structure shown in Figure 3.9b, improves signal integrity; in this case the load structure adds to noise immunity. A common complaint associated with dual rail domino is the resulting increase in area and wiring over single-ended domino. Depending on the architecture, however, true and complement versions of 30% to 80% of all internal nodes may be needed for subsequent logic functions, increasing dual rail efficiency. Unused complementary outputs and their logic trees can be *pruned*, but then preventing continued dual rail domino logic propagation.

In dual rail logic structures such as that shown in Figure 3.9b, summand node precharge is maintained by PFET devices cross-coupled to the opposite outputs. Keeper PFET devices, which depended on inverters, can often be omitted. For local logic operations, the inverter may also be eliminated, and with it, one intrinsic delay and two devices per output. Signals must now be taken from opposite outputs to preserve signal polarity. The buffering supported by the inverter is also gone, so the output fanout load must be watched carefully.

Guaranteed differential outputs also provide opportunities to exploit timing schemes which are not typically possible in single-ended structures. It is known, for example, that the circuit shown in Figure 3.9a has not yet evaluated its inputs if the voltages of nodes N1 and N2 are still both high. A circuit such as a NAND or EXOR, sampling nodes N1 and N2, can sense the fall of one of the two summand nodes and generate a *finish signal*. Finish signals are very useful in qualifying the close of a *transparent latch's* sample period in systems which do not budget for clock skew. Transparent latching and cycle stealing are discussed in Chapter 8. Self-timing is addressed in Section 7.7, "Asynchronous Techniques" on page 281.

3.4.3 Modified Dual-Rail Domino (MDDCVSL)

Modified Dual-Rail Domino (MDDCVSL) Logic is similar to the dual rail domino configurations shown in Figure 3.9a and b, with the exception that the output inverters are omitted to avoid their delay [3.17]. As long as NFET foot switch device 9 is present, there is no *logical* need for the outputs to be inverted[17], and so MDDCVSL does not exhibit true domino action. The required footswitch in each stage, however, reduces the maximum logic depth and capacitive load. The Modified Dual Rail Domino implementation of 2-Way AND logic is shown in Figure 3.9c. Note that it's output node polarities are reversed from those of dual rail domino shown in Figure 3.9a and Figure 3.9b. The output waveforms Q and $\overline{Q}$ are identical to the behavior of nodes N1 and N2 depicted in Figure 3.10.

---

17. That is, there is no reason for the inputs to modified domino circuits coming from other differential dominos to be low during precharge, as long as the footswitch is present.

Once precharged, all inputs to subsequent MDDCVSL stages are high. **Unless *separate* clocks, delayed by cumulative evaluate delay, are used to launch subsequent stage evaluations, the precharge is lost through the enabled footswitch. Modified domino structures are indeed dense and fast, but highly noise sensitive.** These structures do not exhibit true domino effect, and should be considered for use in local logic operations when excess delay can not be eliminated conventional means. Noise is examined in greater detail in Section 4.5, "Noise" on page 157.

3.4.4 Dynamic Differential Split Level Logic (DDSL)

In "Differential Split Level Logic (DSL)" on page 61, a technique which reduced the voltage swing of the summand node to improve performance was described. The pull-up function provided by the cross-coupled latch, shown on page 59, can be fulfilled with clocked PFET precharge devices, as shown in Figure 3.9d [3.18]. This arrangement provides a potential performance improvement by reducing the swing of the precharged node, while still enjoying the performance of precharged logic. In DDSL, precharge/evaluate clock PC performs the same function as in other dual rail structures. When PC transitions low, nodes $\overline{Q}$ and Q are precharged to full rail voltage. With bias V_{REF} set at ($V_{DD}/2+V_{tn}$), the precharge at the top of the logic evaluate tree is limited to $V_{DD}/2$. As with the static structure, DDSL logic depth and output load must be limited.

3.4.5 Pseudo-Clocked Domino

Pseudo-Clocked Domino, like SRCMOS described on page 104, leverages dynamic circuit performance without the need for a global precharge/evaluate clock. Pseudo-Clocked CMOS, however, is not self-timed. Ultimately, it's "clock phase" is determined by preceding logic [3.19]. The pseudo-clocked domino realization of a 2-way AND and NAND is shown in Figure 3.11.

Function

Referring to Figure 3.11, assume logical inputs A and B are outputs of prior dynamic domino circuits. During the prior circuit's precharge interval, A and $\overline{A}$, as well as B and $\overline{B}$ are all low. Subsequently, devices 1-4 are on, and the shown domino stage is also in precharge. Assume A and B, the output of a prior stage, both transition high while $\overline{A}$ and $\overline{B}$ remain low. PFET precharge devices 1 and 2 turn off, ending precharge, and NFET evaluate devices 5 and 7 turn on. Output Q transitions high, and $\overline{Q}$ remains at ground.

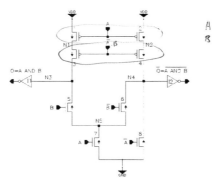

FIGURE 3.11 Pseudo-Clocked Domino

Characteristics

Strengths	Weaknesses
Logically complete	Larger device, area requirements
Reduced clock overhead	Higher fan-out load
High performance	Difficult timing
Self timing and transparent latching options	

TABLE 3-9 Pseudo-clocked Domino attributes.

Pseudo-clocking reduces clock load and sensitivity to clock skew. The area reductions associated with wire and clock regeneration can be substantial. It also increases the fan-out capacitance on the output of prior stages. With this style, precharging **or** evaluation can present the cycle-limiting delay.

Although the clocking function in pseudo-clocked domino has been logically replaced by local signal action, critical timing edges are provided by the first arriving logical input to each stage. Conversely, in conventional domino, the entire path receives the clock simultaneously, and only propagated domino response limits performance. The logical input used as the pseudo-clock should be the earliest arriving of the inputs used in the differential tree in order to avoid collisions between precharge and evaluate, reducing performance by adding discharge current.

3.5 Latched Domino Structures

Sample-Set Differential Logic (SSDL)
Enable/Disable DIfferential Logic (ECDL)
Latch Domino (LDomino)
Differential Current Switch Logic (DCSL)
Switched Output DIfferential Structure (SODS)

The merging of a latch with the load devices of a dynamic logic element provides important features not normally associated with clocked differential logic. These properties are quite useful:

1. The result developed in that stage is fully latched, allowing the circuit to also act as a master or slave latch in a transparently-latched logic scheme[18].

2. Independent of logic content, each stage can be approximated by a unit delay.

3. Higher stage logic content can be realized with the pull-down assistance of a latch.

4. Sense-amplification and additional noise immunity adds design reliability.

5. The latch may save power as it assumes responsibility from the tree for developing output levels. In some structures the tree does not need to complete it's transition swing.

6. The need for an output buffer is eliminated

Latched logic structures, however, are larger and usually slower than their purely dynamic equivalents. The static-like latch adds, at minimum, 4 devices and their associated interconnects. Performance, however, should not be greatly compromised if the available clock is used effectively to avoid hysteresis.

Five structures which incorporate latches enabled by evaluate clocks are shown in Figure 3.12.

3.5.1 Sample-Set Diffential Logic (SSDL)

Sample-Set Differential Logic uses cross-coupling to reduce the voltage swing of the evaluation logic, and to hasten the "decision" delay of the differential evaluate tree. Here, cross-coupling acts as a simple regenerative differential amplifier [3.20], and completes the generation of output begun by the complex tree. A generic representa-

---

18. See Chapter 8.

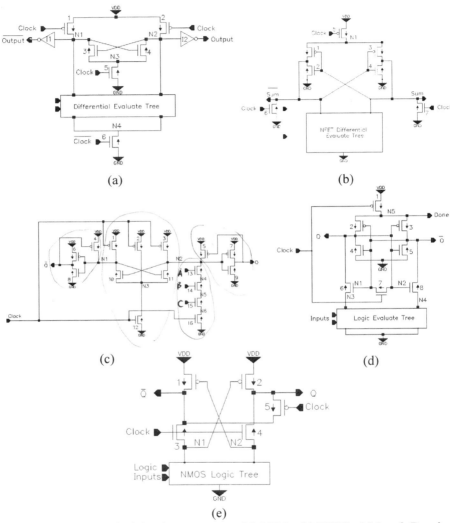

FIGURE 3.12 Latched domino structures: (a) SSDL; (b) ECDL; (c) Latch Domino; (d) DCSL; (e) Switched Output Differential Structure (SODS)

tion of SSDL is shown in Figure 3.12a. Although its structure resembles differential domino, the operation of SSDL is profoundly different.

Function

Complementary logic inputs are provided to the differential evaluate tree in Figure 3.12a. During the "sample" interval, the clock is low, and voltages on node N1 and N2 are set differentially through the current path established by PFET device 1 or 2, the differential evaluate tree, and NFET foot device 6. When the clock transitions high, devices 1,2, and 6 turn off, and differential discharge is arrested. NFET device 5 is on, and the small differential voltage across nodes N1 and N2 is sensed and fully developed by cross-coupled devices 3 and 4. The outputs are inverted and buffered by inverters I1 and I2.

Characteristics

During the sample period, differential voltages are established on the summand nodes by voltage divider action across the resistance of "on" NFET tree devices. During this brief interval, static current establishes these voltages, causing higher power consumption with this style. But since only a small differential voltage needs to be established, the height of the logic tree may increase, device width may decrease, and the delay may be reduced. Rail-to-rail voltages do not need to be achieved: the sense amplifier will quickly take care of the rest during the "set" interval, through a single device to ground. Since correct output depends on accurate sense amplification, channel length and device threshold tracking of devices 3 and 4 are critical. Evaluate nodes N1 and N2 are restored as needed on succeeding cycles. Summand nodes spend virtually no time floating.

Timing of SSDL is simpler than of its DCVS relative, as each stage tends to have the same unit delay. The authors of the original paper describing SSDL report little variation in delay between a circuit 3 devices logically deep compared to that of a circuit 7 devices logically deep. Nonetheless, the sample/set clock frequency must be substantially higher than the system clock, making this style inapplicable in some settings.

3.5.2 Enable/Disable CMOS Differential Logic (ECDL)

Enable/Disable CMOS Differential Logic (ECDL) combines sense amplification and clocked precharging to achieve performance superior to differential domino without consuming the evaluate power of SSDL [3.21] [3.22]. A generic schematic of a clocked ECDL stage is shown is shown in Figure 3.12b.

Function

A given ECDL stage comprises a cross-coupled flip-flop (devices 1-4), a differential logic tree, and resets (devices 5, 6, 7) as shown. The given circuit is said to be *disabled* until Clock transitions low. The low clock *enables,* or powers, the flip-flop via PFET device 5 to latch in the direction determined by the logic tree inputs. Outputs Sum and $\overline{\text{Sum}}$ are permitted to rise, as NFET devices 6 and 7 are off. Typically, the cross-coupled latch achieves its final state well before the differential evaluate tree develops acceptable up and down logic voltages. After the completion of the *enabled* phase, the clock returns high, the output differential voltage collapses, and the ECDL stage is again *disabled.*

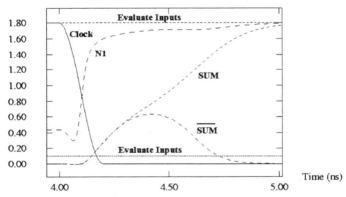

FIGURE 3.13 ECDL Operation

Characteristics

ECDL, like SSDL, uses a cross-coupled sense amp/latch to accelerate the development of output, but avoids the static current SSDL draws during evaluate. The effectiveness of ECDL is strongly dependent on the function implemented. Referenced material [3.22] examines the area and performance trade-offs on ECDL on 4 different adder algorithms. In general, only a few separate circuits are needed in multiple structured instances. Note, however, that in each cycle, one side or the other must be pulled high by latch PFET device 1 or 3, introducing delay dependence on the inferior PFET transconductance.

A **self-timed** realization of ECDL, shown in Figure 3.14, generates "done" signals off the powered latch, which then serves as the clock input to subsequent stages. Local clock synthesis avoids the global clock overhead implicit in clocked ECDL, described previously. The self-timed ECDL stage resembles clocked ECDL, with the addition

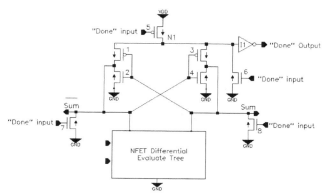

FIGURE 3.14 Self-timed ECDL

of "Done" signal generator/inverter I1, and reset NFET device 6. The DONE signal replaces the clock input to devices 5, 7, and 8 in Figure 3.14. Self-timed ECDL is suited to functions which are iterative, such as adders, where the evaluate path and sequence is determined by the value of the input data. In Chapter 7, the implications of self-timed logic is explored in greater detail.

3.5.3 Latch Domino (LDomino)

As discussed on page 98, single-ended domino is an incomplete logic family; the logic invert function can not be implemented. Latch Domino Logic was offered to provide a circuit which could accept both inverting and non-inverting logic in the first stage of a domino pipeline, allowing use of static as well as dynamic inputs [3.23]. The latched domino realization of the 4-way AND/NAND function is shown in Figure 3.12c.

Function

Latch Domino comprises a single-ended domino tree (devices 13-16), an unbalanced sense amplifier/latch[19] (devices 1-3, 10-12), and half-latched output buffers (devices 5-9). As Clock transitions high, NFET footswitch device 16 turns on, developing on

---

19. The sense latch is said to be unbalanced due to the difference in capacitance on nodes N1 and N2

---

node N2 the logic state defined by inputs A-C. NFET device 12 also turns on, allowing cross-coupled NFET latch devices 10 and 11 to both conduct. If the logic tree pulls node N2 to ground, the cross-coupling action of devices 10 and 11 retains a high level on Node N1. If the logic tree remains high, the latch's load imbalance causes node N1's voltage to dip when device 12 initially turns on, enough to flip the latch. With the latch set in either direction, devices 5-9 invert and buffer the complementary outputs.

Characteristics

Latched domino logic accepts static or dynamic inputs, single ended or dual-rail, making useful as the first interface stage in a dynamic logic path. The structure can execute complex logic. Its higher device count, however, is less than that of the entire complementary logic function needed to recreate the complement output, but has additional delay. Besides inverting the logic, the latch acts as a regenerative amplifier, which can improve standard domino or DCVS performance by assisting in the pull-down of a loaded summand node. Charge redistribution[20] may also not be as severe in LDomino, as the logic inputs may not be monotonic: intermediate nodes such as N4-N6 may have been precharged. True and complement outputs are symmetrically inverted and buffered.

LDomino also compromises specific characteristics. At the rise of the evaluate clock, both sides of the latch initially conduct until the latch sets, eroding precharged noise margin. The latch, then, must be highly *imbalanced*. Noise margin is improved by beginning evaluation before the latch sets. Latch-set delay, however, has device count and performance penalties; quick settling increases the latch's discharge assistance to the logic tree and improves noise immunity, but increases area and power consumption. Also, true and complement output delays differ, and must be anticipated. Input glitching is a concern with this design style, but may be avoided by stabilizing inputs before evaluation begins. Latched domino's complexity suggests effective use in selected settings.

3.5.4 Differential Current Switch Logic (DCSL)

Power consumption has been a deterrent to using differential structures; one of the two sides is certain to discharge and require recharge every cycle. In addition, although NFET evaluate trees achieve efficiency by growing logical width and depth, each additional device level increases junction capacitance and power. Differential

---

20. See "Capacitive Charge Divider," page 96

Current Switch Logic (DCSL), shown in Figure 3.12d, achieves high performance as well as lower power in clocked applications by reducing voltage swing [3.24].

Function

Referring to Figure 3.12d, assume the Clock input is high, and that the given circuit is in *"pre-discharge"* mode[21]. Device 7 is on, causing devices 6 and 8 to discharge outputs Q and $\overline{Q}$ to a voltage of less than one NFET threshold voltage. When Clock goes low, the circuit is said to enter *evaluate* mode. PFET device 1 turns on, and PFET devices 2 and 3, both in the on-state, provide drive for Q or $\overline{Q}$ high. A path from ground to node N3 or N4 prevents N1 or N2 from rising, decoupling the evaluate tree from the opposite side and allowing it to rise quickly.

Characteristics

The previous cross-coupled DCVS circuits shown in this section employ the sense amp/latch structure to assist the NFET evaluate tree in achieving rail-voltage signals quickly. DCSL does the opposite. As soon as the differential sense-amplifier achieves a sufficient signal to develop an output state, it *disconnects* the evaluate tree from the latch, allowing the high latch side to set unencumbered. The disconnection not only reduces the delay in developing the output, but also reduces the voltage swing, and hence power, in the evaluate tree. In addition, since the logical inputs no longer couple to the high output after the evaluate has occurred, the opportunity for transient DC currents, or for noise to disturb the outputs is limited to the interval preceding latching. During this interval, the circuit is exquisitely sensitive, given the high amplifier gain. Finally, on nodes of significance, only full rail voltages are developed, avoiding problems associated with partially-on devices, found in other styles. As with ECDL, a "Done" signal may be formed as the latch is powered, DCSL to **self-time** a logical path. Self-timing is simplified by the decoupled nature of DCSL inputs: inputs do not need to remain valid after "set" to retain output data.

3.5.5 Switched Output Differential Structure (SODS)

ECDL improved upon static DCVS by introducing a latch gated by a control input, resulting in the elimination of output buffers, and considerable area and power savings. Switched Output Differential Structures (SODS) enhances ECDL performance

21. DCSL preconditioning, achieved by discharging to voltage V_{tn}.enables the same action accomplished by the precharging to V_{DD} practiced by other styles discussed.

improvements while providing static-like reliability. A logic clock is used to avoid contention issues [3.25]. SODS is generically represented in Figure 3.12e.

Function

Referring to Figure 3.12e, when the system input "Clock" is low, NFET devices 3 and 4 are off and prevent DC power loss; the structure is said to be in precharge mode. PFET device 5 is on, tying the voltages of $\overline{Q}$ and Q to each other. Because in SODS signals remain complementary during precharge, node N1 or N2 is grounded, turning on PFET precharge device 1 or 2. Device 5 then provides a charging path to the remaining node. When Clock transitions high, the already-stable NMOS evaluate tree differentially sets the latched voltage on node $\overline{Q}$ and Q through isolation NFET devices 3 and 4. The enabled regenerative feedback of devices 1-4 hastens the setting of valid outputs.

Characteristics

SODS "gets it both ways." The structure effectively uses a clock input to achieve dynamic performance, yet at no times requires its outputs to float. Designing to distinct precharge and evaluate intervals allows the use of optimized device sizes; PFET precharge devices 1 and 2, for example, are removed from the critical evaluate path, and have a fixed precharge interval   Output capacitance is reduced, compared to differential domino[22] or ECDL. The trade-off, however, is limited logical depth, as the access to ground is through the evaluate tree, and access to precharge is through 2 PFETs.

3.6 Clocked Pass-Gate Logic

Dynamic Complementary Pass Gate Logic
Sense-Amplifying Pass-Transistor Logic

The relationship of clocked pass gate structures to static pass structures described in Section 2.4, "Non-Clocked Pass-Gate Families" on page 65, is analogous to that of dynamic domino logic and static logic. Clocked pass-gate-based structures require preconditioning, and are operated in a monotonic fashion. So similarly, this dynamic logic style embodiment is vulnerable to the charge loss mechanisms described on

---

22. See Section 3.4.1, Differential Domino on page 111

page 96 in addition to the basic pass gate concerns outlined in Section 2.4 on page 65. Two design styles representative of clocked pass-gate structures are described below.

3.6.1 Dynamic Complementary Pass Gate Logic (DCPL)

Dynamic Complementary Pass-Gate Logic (DCPL) incorporates favorable attributes of Domino CMOS[23] and CPL[24] styles [3.26]. Both true and complement inputs must be provided to clocked pass-gate structures, and *must transition monotonically in a polarity determined by their connection.* The clocked pass-gate realization of the 2-WAY AND logic function is shown in Figure 3.15.

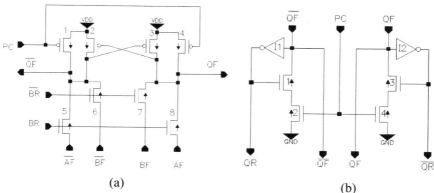

(a)   (b)

FIGURE 3.15 Dynamic Complementary Pass-Gate Logic; structure executing 2-way AND (a); buffer/inverter adding rising outputs, passing falling outputs (b).

Function

Referring to Figure 3.15a, clock PC is at ground during precharge, allowing PFET devices 1 and 4 to precharge potentially falling output nodes QF and $\overline{QF}$ to voltage V_{DD}. The logical inputs to sources of DCPL pass devices *must be at voltage V_{DD} during precharge*, and high or falling, in a complementary, differential manner, during evaluate. They can be taken directly from nodes QF and $\overline{QF}$ of a prior stage. Referring to Figure 3.15b, during precharge, potentially rising inputs QR and $\overline{QR}$ are predischarged by NFET devices 2 and 4. All logical inputs to the gates of DCPL pass

---

23. See Section 3.2.1, Domino CMOS on page 93
24. See Section 2.4.3, Complementary Pass Gate Logic (CPL) on page 73

---

devices *must be low during precharge*, and low or rising in a complementary, differential manner during evaluate. These inputs are *inverted* DCPL outputs from a preceding stage. Assume the logical inputs shown in Table 10 below are provided to the DCPL circuit shown in Figure 3.15.

Phase \ Input	AF	$\overline{\text{AF}}$	BR	$\overline{\text{BR}}$	BF	$\overline{\text{BF}}$
Precharge	1	1	0	0	1	1
Evaluate	1	0	1	0	1	0

TABLE 3-10 **Logical inputs for example DCPL operation**

With the indicated inputs, NFET pass gate devices 5 and 8 are turned on at the onset of evaluate, coupling node $\overline{\text{AF}}$'s low state and AF's high state to nodes $\overline{\text{QF}}$ and QF, respectively, achieving the function QF=AB.

Characteristics

Strengths	Weaknesses
NFET evaluation devices only	Rising & falling complementary I/O needed
Monotonic, glitch-free evaluation	Added clock loading of buffer
Built-in logical sequencing of path stages	Charge redistribution sensitivity

TABLE 3-11 **DCPL attributes.**

The benefit of generating rising and falling, true and complement signals is found in the elimination of low-to-high transitions through NFET evaluate devices in the critical path. The structure is quite frugal in device use, and provides a dynamic performance enhancement to static CPL logic. DCPL rising outputs always trail falling outputs by at least one inversion delay (inverter I1 or I2 in Figure 3.15b), implying that the path delay of a sequence of DCPL structures is controlled by the gate inputs to devices in source-stabilized evaluate trees.

Optional cross-coupled PFET devices 1 and 2 are present only to enhance noise immunity. With static pass gate styles, the PFET devices were essential, to restore full voltage to the summand.

In addition to the hazards of pass gates described in Section 2.4 on page 65, precharged pass gates are vulnerable to precharge or performance loss caused by Miller effect coupling of input transitions to the pass-gate to the precharged true or complement output. Pass gate structures in general require thorough usage checks; operating a pass-gate structure dynamically requires further vigilance.

3.6.2 Sense-Amplifying Pipeline Flip-Flop (SA-F/F) Scheme

In the previous dynamic pass gate strategy, complexity was accepted in order to guarantee that only full rail voltages would be developed, and that threshold voltage drops would never be experienced on outputs. Logic built with the SA-F/F scheme intentionally invokes threshold voltage drop to reduce delay [3.27] [3.28]. The SA-F/F structure providing the XOR function is shown in Figure 3.16 below.

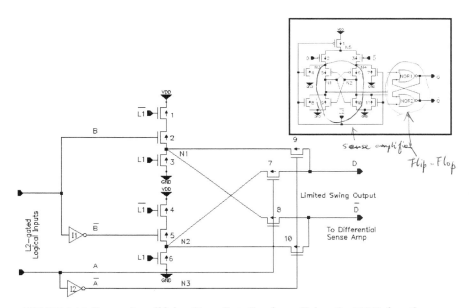

FIGURE 3.16 Sense-Amplifying Pass-Gate Logic realizing the XOR function. Insert illustrates accompanying differential sense amplifier/Flip-Flop.

Function

Clocks L1, $\overline{L1}$, L2, and $\overline{L2}$ are provided to the circuit. Logical inputs A and B are gated to the circuit block by $\overline{L2}$ (not shown); complement versions of each are then developed by inverters I1 and I2. Clock input L1, when high, preconditions nodes N1 and N2 both to ground. As $\overline{L1}$ rises and L1 falls, NFET device 2 or 5 turns on, depending on the stabilized value of B, and provides a differential voltage to nodes N1 and N2. Note that since conditioning devices 1-6 are all NFETs, voltage excursions on node N1 and N2 will never reach voltage V_{DD}. After briefly developing differential between outputs D and $\overline{D}$, clock L2 falls, enabling the Renshaw-type sense amplifier/latch [3.29] shown in the inset of Figure 3.16 to generate full rail differential outputs Q and $\overline{Q}$.

Characteristics

Strengths	Weaknesses
Small logical voltage swings	Terminating logic block in path only
Low stage propagation delay	Efficient only for heavily loaded fan-out

TABLE 3-12 SA F/F scheme attributes.

SA-F/F recreates master/slave latch action, with devices 1-10 forming the master, and the sense amplifier (shown in the inset) forming the slave. Since the sense amplifier/latch synchronizes output to the system, a race condition easily induced by self-timing the sample period of the sense-amplifier[25] is avoided. The clock frequency (incidence of L2 falling edge controlling the sense-amplifier sample period) is simply increased until the maximum achievable evaluation rate is exceeded, in a scenario where L1 is generated off the opposite rising edge of L2. Because L2 is actually the system clock, and because L1 is derived from that same clock, the structure has no built-in race, and adapts well to single-phase-clocked architectures[26].

A SA-F/F stage contains the latch terminating the logic path it participates in. Hence, rather than an independent circuit design style, this structure accepts the full rail output of most of the logic topographies previously discussed. The complexity and device/area overhead associated with SA-F/F must be amortized by substantial logic content. For small logic evaluations, it is quite inefficient and easily surpassed in performance and area by other styles. Stages that receive slow, heavily loaded inputs, or stages with high logic width and depth are both good candidates for SA-F/F path termination.

3.7 Summary

A variety of single-ended and dual rail domino, latched, and pass gate circuit families appearing in the literature demonstrate the performance advantage of accommodating only monotonic inputs. Circuits designed for monotonic operation have removed

---

25. Sense amplifier self-timing is common in arrays, where dummy or reference word and bit lines predict signal delay and provide a delay line for the strobing of the sense amplifier.

26. See Chapter 7.

---

from their evaluate path many of those elements which increase delay and power consumption, such as high fan out capacitance, latch hysteresis, and cross-over current. The overhead, although removed from the critical path delay, is larger than that found in static circuits. This overhead may include clocks, sense-amplifiers, output voltage conditioners, or latches. A number of innovative structures shown select different mixes of noise immunity, performance, power consumption, and area.

From Section 1.3, "Charge Loss Mechanisms" on page 25, we were prepared for *process-induced* sensitivities in dynamic circuits which were not present in static designs. In this chapter, clocked structures were shown to carry *design-induced* sensitivities as well. In static designs, design and process sensitivities generally impacted performance and yield; in clocked designs, basic circuit functionality can be lost.

In the next chapter, we examine the *performance variation* caused by our choice of design style, process, and application conditions. Insight into the sources of performance variability allows informed decisions when budgeting timing and voltage margin.

REFERENCES

[3.1]   T. Williams, "Dynamic Logic: Clocked and Asynchronous," *1996 IEEE International Solid State Circuits Conference Tutorial #4.*

[3.2]   R. H. Krambeck, etal., "High Speed Compact Circuits with CMOS," *IEEE Journal of Solid-State Circuits*, Vol. SC-17, No. 3, June 1982, pp. 614-619.

[3.3]   I. S. Wang, etal., "Ultrafast Compact 32-bit CMOS Adders in Multiple-Output Domino Logic," *IEEE Journal of Solid-State Circuits*, Vol. 24, No. 2, April, 1989, pp. 358-369.

[3.4]   Z. Wang, etal., "Fast Adders Using Enhanced Multiple-Output Domino Logic," *IEEE Journal of Solid-State Circuits*, Vol. 32, No. 2, February 1997, pp. 206-214.

[3.5]   T.W. Houston, etal., "Compound Domino CMOS Circuit," *U.S. Patent #5015882*, May 14, 1991.

[3.6]   F. Murabayashi, etal., "2.5V Novel CMOS Circuit Techniques for a 150 MHz Superscalar RISC Processor," *Proceedings of the 21st European Solid-State Circuits Conference/IRC-95*, pp. 178-181.

[3.7]   H. Yamada, etal., "A 13.3 ns Double-precision Floating Point ALU and Multiplier," *Proceedings of the 1995 International Conference on Computer Design (ICCD)*, pp. 466-470.

[3.8]   G. Yee, etal., "Clock-Delayed Domino for Adder and Combinational Logic Design," *Proceedings of the 1996 International Conference on Computer Design*, pp. 332-337.

[3.9]   J. Silberman, etal., "A 1.0GHz Single-Issue 64b PowerPC Integer Processor," *Proceedings of the 1998 IEEE International Solid-State Circuits Conference*, pp. 230-231.

[3.10] W.-H. Lien, W. Burleson, "Wave-Domino Logic: Theory and Application," *Proceedings of International Symposium on Circuit and Systems*, 1992, pp. 2949-2952.

[3.11] A. K. Woo, etal., "Static PLA or ROM circuit with self-generated precharge," *U.S. Patent #4728827*, December 3, 1986.

[3.12] R. A. Haring, etal., "Self Resetting Logic Register and Incrementer," *1996 Symposium on VLSI Circuits, Digest of Technical Papers*, 1996, pp. 18-24.

[3.13] M. K. Ciraula, etal., "Self-Restore Circuit with Soft Error Protection for Dynamic Logic Circuits," *U.S. Patent #5,706,237*, January 6, 1998.

[3.14] N. Goncalves, etal., "NORA: A Racefree Dynamic CMOS Technique for Pipelined Logic Structures," *IEEE Journal of Solid-State Circuits*, Vol. SC-18, No. 3, June 1983, pp. 261-266.

[3.15] C.M. Lee, etal., "Zipper CMOS," *IEEE Circuits and Devices Magazine*, May 1986, pp. 10-16.

[3.16] L. G. Heller, etal., "Cascode Voltage Switch Logic: A Differential CMOS Logic Family," *Proceedings of 1984 IEEE International Solid-State Circuits Conference*, pp. 16-17.

[3.17] P. Ng, etal., "Performance of CMOS Differential Circuits," *IEEE Journal of Solid-State Circuits*, Vol. 31, No. 6, June, 1996, pp. 841-846

[3.18] Leo C. M. G. Pfennings, etal., "Differential Split Level CMOS Logic for Subnanosecond Speeds," *IEEE Journal of Solid-State Circuits*, Vol. SC-20, No. 5, October, 1985, pp. 1050-1055.

[3.19]   E.Miersch, etal., "Pseudo-clocked Cascode Voltage Switch Logic System," *IBM Technical Disclosure Bulletin*, Vol 28, No. 6, November, 1985

[3.20]   T. A. Grotjohn, etal. "Sample-Set Differential Logic (SSDL) for Complex High-Speed VLSI," *IEEE Journal of Solid-State Circuits*, Vol. SC-21, No. 2, April, 1986, pp. 367-369

[3.21]   Shih-Lien Lu, "Implementation of Iterative Networks with CMOS Differential Logic," *IEEE Journal of Solid-State Circuits*, Vol. 23, No. 4, August, 1988, pp. 1013-1017

[3.22]   Shih-Lien Lu, etal. "Evaluation of Two-Summand Addrs Implemented in ECDL CMOS Differential Logic," *IEEE Journal of Solid-State Circuits*, Vol. 26, No. 8, August, 1991, pp. 1152-1160.

[3.23]   J. Pretorius, etal., "Latched Domino CMOS Logic," *IEEE Journal of Solid-State Circuits*, Vol. 21, No. 4, August, 1986, pp. 514-522.

[3.24]   D. Somasekhar, etal., "Differential Current Switch Logic: A Low Power DCVS Logic Family," *IEEE Journal of Solid-State Circuits*, Vol. 31, No. 7, July, 1996, pp. 981-991.

[3.25]   A. J. Acosta, etal., "SODS: A New CMOS Differential-Type Structure," *IEEE Journal of Solid-State Circuits*, Vol. 30, No. 7, July, 1995, pp. 835-838.

[3.26]   I. Dobbelaere, "Dynamic Complementary Pass-Transistor Logic Circuit," *U.S. Patent # 5399921*, March 21, 1995

[3.27]   M. Matsui, etal., 200MHz Video Compression Macrocells Using Low-Swing Differential Logic," *Proceedings of 1994 IEEE International Solid-State Circuits Conference*, pp. 76-77, 314.

[3.28]   M. Matsui, etal., "A 200 MHz 13 mm2 2-D DCT Macrocell Using Sense-Amplifying Pipeline Flip-Flop Scheme," *IEEE Journal of Solid-State Circuits*, Vol. 29, No. 12, December 1994, pp. 1482-1490

[3.29]   D. Renshaw, etal., "Race-Free Clocking of CMOS Pipelines Using a Single Global Clock," *IEEE Journal of Solid-State Circuits*, Vol. 25, No. 3, June 1990, pp. 766-769.

CHAPTER 4   *Circuit Design Margin and Design Variability*

4.1 Introduction

In the preceding chapters, process variations and circuits styles were discussed. Each circuit style has its own reaction to variations of the process. Each variation must be accounted for to maintain the functionality and desired speed of the circuit across these distributions. All process parameter distributions are a function of the range that the parameter is critical both spatially and temporally. This chapter will investigate the variation of the process on static CMOS logic, dynamic domino, pass gate and DCVS logic.

Variations of process, power supply, and temperature are important in two different forms. First is the variation from wafer to wafer if all parameters were exactly controlled. This global variation would lead to a change in performance between two different wafers, but each circuit within a chip would be identical. A similar situation is true for the supply voltage and temperature. The second form is localized where the process, power supply and temperature vary across the chip. In this case, a given circuit, though design the same as another circuit, will have a different performance capability.

A specific design will introduce variation upon itself due to a particular circuit style chosen, layout configuration for transistors and interconnects and signal flow. As a designer, this is the area that one has the most impact on maintaining the functionality of circuit. Noise within a chip is one area of design induce variations and a major contributor to its functionality. One must consider the induced noise upon a circuit due to interlayer/intralayer coupling on the metal interconnects. Self imposed noise is an issue for large drivers that drive a transmission line and modeling the capacitance as if it is distributed versus a lumped element capacitance. All noise issues are increased when switching of neighboring circuits is done simultaneously and creating variations upon the power supply. Gradients of temperature across a chip contribute to variations in delay.

Circuits that fail for reason other than timing may have a root cause in the leakage of the process, noise within the system, or soft error fails from alpha particles.

Taking these issue into account, this chapter will consider the overall budget required for timing the slowest paths on a chip (slow paths) and timing the race conditions (fast paths) and creating a satisfactory noise margin in a CMOS circuits.

4.2 Process Induced Variation

For this discussion, the process variables considered are the variations in the front end of line (FEOL) including effective length (L_{EFF}), effective width (W_{EFF}), threshold voltage (V_T), the oxide thickness (T_{ox}), leakage, Miller capacitance, hot electron carrier and the back end of line BEOL including interconnect resistance and interconnect capacitance. As mentioned in Chapter 1 the variability of each of these parameters have a different range of variability and are not completely independent of each other.

4.2.1 Static CMOS Logic

Static CMOS operation must quickly drive moderate capacitive loads, and so has a strong performance dependence on parameters which enhance I_{DS}, such as channel length, V_{DD}, threshold voltage, and junction temperature. An accurate delay model for a static circuit must anticipate process and application condition variation. As described in Section 1.2.1 short channel effect must be anticipated.

Static CMOS is also a strongly ratioed logic style and is very resilient at maintaining functionality across all possible process and application corners. The major variability is the performance, noise margins and switch points associated with the PFET-to-NFET relative strength. Strong tracking of the two device types is important for good yield and reliability. Table 1 shows the major areas of concern for static CMOS circuits. The process portions will be dealt with in this section and the other areas will be dealt with in there respective sections.

Static CMOS Circuit Sensitivities

L_{EFF} Variation	W_{EFF} Variation	V_T Variation	V_T, L_{EFF} Tracking	N/P V_T, L_{EFF} Tracking	Substrate Sensitivity	Leakage	Hot Carrier Wearout	Interconnect Resistance	Interconnect Capacitance	Logic Noise	Delay Noise	Supply Noise	Soft Errors (alpha,gamma)	Temperature	Supply Tolerance

TABLE 4-1   Static Combinatorial CMOS Logic Sensitivities

If the process parameters are maintained within their given parameters, static circuits will not fail. However, the variation in delay is subject to each parameter. Consider two timing paths consisting of inverters, NANDs and NORs. One timing path does not have any significant RC delay and another will have 20% of its total delay as RC delay. If the process parameters independently varied by three sigma, the change in delay is shown in Figure 4.1. Included in this figure is the performance response to a 50 degree temperature range and an eight percent voltage tolerance. No device mismatching was considered in these simulations.

The largest variation in performance of static CMOS circuits is achieved by varying the L_{EFF} of the FET. As expected, the path with 20% RC delay does not vary as much because a smaller percentage of the timing path is dominated by the FET response to L_{EFF}. The oxide thickness, temperature dependence and the voltage dependence all vary the saturation current by an equivalent amount, so there delay variations were nearly identical. All three parameters also vary the mobility of the majority carrier and threshold voltage of the FET.

W_{EFF} is rarely a parameter to consider since most devices are physically long enough that the ΔW is a small percentage of the total gate width. If the design is using very narrow devices, then the narrow channel effect should be considered. For example, caches often use very narrow devices in the six transistor cell for both the PFETs and the pass gate NFETs controlled by the word line.

Surprisingly, the variability of the threshold voltage did not induce much variation on the performance. If the threshold voltage is tightly controlled, then this would be the case. Consider the saturation current equation, $I_{DS} = 0.5\beta(V_{GS} - V_T)^2$. If the V_T is 15% of the V_{GS} and V_T varies by 10%, then the change in saturation current is less than 4%. The process cannot reduce the threshold voltage blindly without considering the effects of subthreshold leakage of an FET, since the subthreshold leakage current increases a decade every 60-80mV decrease of the V_T. A lower V_T will increase the DC current consumed. This is a major issue for low power circuits and circuits used in mobile applications with short battery life. Increased leakage does not inhibit the functionality of static circuits, but is a large concern for dynamic circuits (see Section 4.2.2 on page 139).

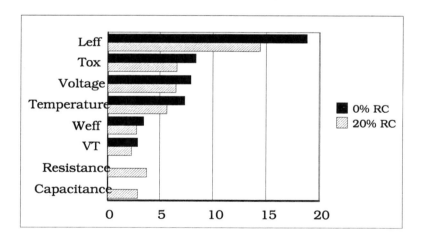

FIGURE 4.1 Delay variations of a static CMOS circuit with process and environment.

The variation in the interconnect resistance and capacitance affect both the delay of the driving static circuit and RC delay down the interconnect. Resistance is a function of the wire thickness, resistivity and width. Capacitance is a function of wire thickness, width, intralayer space, interlayer dielectric thickness and permittivity of the dielectric. Capacitive coupling will be discussed in Section 4.5.1 . A larger interconnect resistance increases the delay of the RC component. It also can have the effect of speeding up the output node of the driving circuit since the interconnect resistance shields the driving circuit from the capacitances at the far end of the wire. This results in the rise and fall times at the output of the driving circuit being much faster than the far end of the wire. This widening of the pulse as it travels down a wire is known as dispersion. In most timing models, this is handled by treating the load on the driving circuit as an effective capacitance that is less than the total capacitance [4.1] . This implies that there is a limit to the usefulness of increasing the size of a driving circuit when the wire has a large amount of resistance near the driver.

The switch point ($V_{IN} = V_{OUT}$) of a static CMOS inverter is a function of the design dimensions and the process variables. Typically one designs an inverter for maximum speed or equal rise and fall times on the output. This results in a ratio of the width of the PFET to the NFET (P/N) in the range of 1:1 to 2:1.

> **Rule of thumb:** Static CMOS P/N width ratio is nearly 2.0 for equal rise and fall times and nearly 1.5 for maximum speed.

Figure 4.2 shows the switch point for a range of P/N width ratios across the process and environment. The nominal process is the center line. If one transition is known to be more speed critical than the other, then one may design the with P/N width ratio skewed away from 2/1. For example, to speed up the rising transition into an inverter, then choosing a larger smaller P/N ratio will reduce the switch point and allow the gate to transition sooner. Moving the switch point away from $V_{DD}/2$ has a trade-off with noise margin. The smaller one adjusts the P/N width ratio, the lower the switch point is for the inverter. At the same time, the unity gain point for the rising input is reduced. For static CMOS where the power supplies are considered ideal, the unity gain point is equal to the amount of noise margin [4.2] . Therefore, one has reduced the amount of upward noise acceptable on the input of the inverter. At the same time, the noise margin for the falling input transition has been increased. A similar situation exists when increasing the P/N width ratio to speed up a falling input transition. The noise margin for a falling noise event from V_{DD} is reduced, while the upward noise margin has been increased.

> **Rule of thumb:** It is ill advised to design outside of the P/N ratio range of 1/2 to 4/1 because the noise margin will be degraded.

The Process Max and Process Min data in Figure 4.2 includes all variations across a process within three sigma. The largest portion of the process variation component for the switch board variation is the mistracking of the NFET L_{EFF} to the PFET L_{EFF}. Mistracking as the L_{EFF} is shorter is a larger problem since the short channel effect induces a larger delta in the threshold voltage due to the roll-off at these shorter channels as shown in Figure 1.9 on page 14. The switching threshold of a CMOS inverter will vary by nearly 20% when the two devices are mismatched. This variation will change the delay of a circuit by speeding up one transition of an inverter, but decreasing the speed of the opposite transition. If two static circuits are in series with each other and have the same mistracking between the PFET and NFET, then the same mismatch in delay on each inverter will approximately cancel each other and the delay will be close to the matched case. Over a long path with many different gates, the variation due to ACLV will average out and approach a nominal situation.

Total Max and Total Min data in Figure 4.2 includes the process variations plus the variations due to external V_{DD} and temperature ranges. This is the total range that one needs to design to.

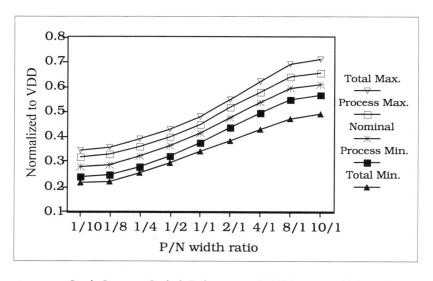

FIGURE 4.2 Static Inverter Switch Point versus PFET/NFET Width Ratio.

Hot electron degradation is a component of variability on static circuits. Since hot electron effect is most likely to happen when the current is flowing with high transverse electrical fields in the device, the most susceptible static CMOS logic gates are those with slow input transitions and with high output loadings. Static circuits do not have a DC current component to compound the hot electron degradation. Also, hot electron degradation causes more performance shifting in bidirectional circuits and static circuits only carry the current in a unidirectional manner. The timing of two gates with different loadings will start to diverge with time as the number of electrons trapped in the oxide increases in the heavily loaded gate. This mechanism is a function of number of times that the circuit has switched, the conditions during the switch and the previous amount of degradation.

Miller capacitance is a parasitic capacitance due to the overlap of the gate with the source and drain diffusion regions as in Figure 1.12 on page 19. This capacitance is connected directly from the input to the output of an inverter. Therefore, a fast rising input will supply charge through this capacitor onto the output node raising the output node above V_{DD}. This effect is called the Miller effect or "bootstrapping" and can be seen by the rise in signal N1 above V_{DD} in Figure 2.3 on page 56. The Miller effect will require the NFET to pull the output node down from a value slightly larger than V_{DD}. This additional ΔV will slow down the performance of the inverter by the percentage with which the output node was raised above V_{DD}. This percentage is determined by the ratio of the amount of Miller capacitance to the amount of capacitance on the output node of the inverter. The voltage increase above V_{DD} cannot be greater than the forward bias voltage of the PN diode formed by the drain of the PFET and the NWELL that the PFET resides in. This diode will clamp the amount of Miller effect. Also, the Miller effect is not realized if the input slope slew rate is too slow [4.10].

The Miller capacitance will cause the input capacitance on the gate of an FET to appear as a larger capacitance since the charge supplied to the input is now distributed to charging the gate and charging the drain. The change in the input capacitance is a function of the transconductance of the transistor [4.3] . Noise associated with Miller capacitance will be covered in Section 4.5

4.2.2 Dynamic Domino Logic

Dynamic domino logic was introduced in Section 3.2.1. For this discussion, reference will be made to the dynamic domino NAND structure as shown in Figure 3.1 on page 94. When necessary, discussion of a NOR structure will be used. All discussions will be made without reference to the optional foot switch used in some dynamic domino logic circuits. Dynamic domino circuits are employed when static CMOS circuits would not produce the speed necessary. This circuit family comes with its own set of hazards.

This section will discuss the process variability for performance and Section 4.5 will address the issues of noise. Table 2 highlights the major area of concern with circuit sensitivities. We have added the V_T variation and leakage as an area of concern. By definition, the dynamic domino does not have a logical equivalence of PFETs for the logical evaluation on the dynamic node one can remove the sensitivities to the N/P V_T, L_{EFF} tracking for that portion. Also, by definition of dynamic domino, an inverter (and sometimes a NAND or NOR gate) is places at the output of the dynamic node to make the circuit non-inverting. Therefore, all statements in the last section on static logic apply to this driving output inverter.

Dynamic Domino Circuit Family Sensitivities

L_{EFF} Variation	W_{EFF} Variation	V_T Variation	V_T, L_{EFF} Tracking	N/P V_T, L_{EFF} Tracking	Substrate Sensitivity	Leakage	Hot Carrier Wearout	Interconnect Resistance	Interconnect Capacitance	Logic Noise	Delay Noise	Supply Noise	Soft Errors (alpha,gamma)	Temperature	Supply Tolerance

TABLE 4-2 Dynamic Domino Circuit Sensitivities

To attain the best performance from a dynamic domino circuit, the goal is to discharge the dynamic node N1 as fast as possible and then drive the signal to the next circuit through the output inverter. To discharge N1 quickly, the design parameters are the size of the NFET stack, the switch point of the output inverter and the size of the PFET keeper. The PFET precharge device has little to do with the performance of the circuit. Its only responsibility is to initially charge the dynamic node to V_{DD} within the time allotted for the precharge portion of the cycle.

Increasing the size of the NFETs 1 and 2 in Figure 3.1 will increase the speed of this particular circuit since the ability to discharge N1 is increased, but doing this has increased the burden on the previous circuit driving the inputs. The NFET stack should be sized to quickly drive the output inverter. Two separate switch points exist with this non-inverting style. First, the switch point for node N1 is nearly equal to the threshold voltage of the NFET in the stack. This being the case, the input only needs to rise to V_{TN} to start the node from discharging. This is much lower than the switch point of $V_{DD}/2$ for the static logic. Herein lies the speed advantage. The second switch point is the output inverter. With the critical timing being the dynamic node N1 falling, a case is provided for skewing the P/N width ratio of the output driver to a larger value such as 4/1 or 8/1, remembering that we are again sacrificing noise margin.

With the switch point of the inverter of the input reduced, speed as a function of the variation of V_T is much more sensitive, however, L_{EFF} is still the most significant contributor to process variation. A one sigma change in L_{EFF} will again result in about a 5-6% performance difference. T_{ox}, V_{DD} and Temperature all effect V_T, therefore they are still large contributors to the delay.

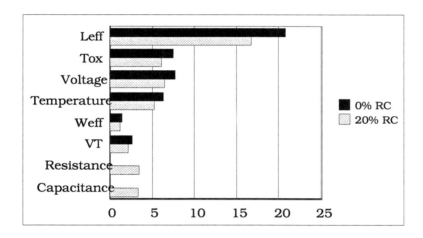

FIGURE 4.3 Delay variations of a dynamic domino CMOS circuit with process and environment.

If one could control the noise, why not allow the V_T to be very small? The answer is leakage! Since subthreshold leaking increases substantially with decreasing threshold voltage, leakage will start to discharge the dynamic node. This will cause a fail because it is not recoverable. If the dynamic circuit is a wide NOR structure without a footer device, the leakage is a sum of the width of all the NFETs times the subthreshold leakage current. To prevent leakage from disturbing the dynamic node, a weak PFET (Device 2 in Figure 3.2.1 on page 93) is used for maintaining the precharge. This structure is called a half-latch since it maintains only a precharge node and does not hold a discharged node. The PFET is used for overcoming leakage and for added noise margin. It will slow the circuit down since the NFET stack must now overcome the saturation current of the PFET in addition to the capacitance on the node. One would not want to increase the feedback strength too much. Figure 4.4

shows how the delay of a NAND structure varies as a function of the PFET's width. Small PFETs (W ~ 1μm) have little impact to the overall delay, while large PFETs will hurt the delay and eventually the NFET stack will not overcome the PFET strength and the dynamic node will never be discharged.

Rule of thumb: Design the keeper to supply 5-10X more than the worst case leakage. This will guarantee to prevent any leakage concerns and will not provide much decrease in speed.

Rule of thumb: The effective NFET stack strength should be 3-5X the strength of the keeper to maintain functionality.

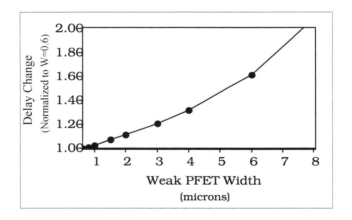

FIGURE 4.4 Delay vs. the weak PFET width.

W_{EFF} variation is not a concern with dynamic circuits since all widths are not near the minimum. Even the weak PFET keeper is not a concern since there is a lot of margin in the design using the Rule of thumb stated above.

Hot electron effect is the same as it was for static. The NFETs that pull the dynamic node down will not be in their linear region very long, so little degradation is expected. Also, there is typically not much capacitance on the dynamic node.

Miller capacitance is more important in dynamic circuits than static circuits. This is due to the "dynamic" nature of the internal node. The "bootstrapping" mechanism is the same as described before, but this there is less capacitance on the internal node since the equivalent PFETs are not present. Without these equivalent PFETs, the weak PFET "keeper" must be able to accommodate any charge injected onto the dynamic node across the Miller capacitance.

Interconnect variations is very similar to the static case if the dynamic node is not too long, because the interconnect is driven by a static inverter. When the dynamic node is long, the weak keeper must not let the resistance of the wire reduce it's effectiveness at holding the far end of the dynamic node high. Also, long dynamic nodes have more potential for other wires coupling to them and creating noise. The variation in capacitance on a long dynamic node will vary the timing greatly and is not reflected in Figure 4.3.

Other dynamic circuit styles like Multiple-Output Domino Logic, Compound Domino Logic, Self Resetting CMOS or Clock-Delayed Domino Logic will respond to process variations much like the Dynamic Domino circuits. The Noise Tolerant Precharge circuits will behave like dynamic circuits with a a skewed switch point due to the complementary PFET logic. This switch point would act like a static circuit that has a PFET width to NFET width ratio of 10:1. The alternating styles of logic like NORA Dynamic and Zipper Domino will have additional process constraints due to the variations of the V_T of the PFETs in the alternating pullup logic trees.

4.2.3 Pass Gate Logic

Pass gate circuits such as Figure 2.7 on page 66 are much more tightly coupled to the process than any other circuits. They have the most performance variations and the largest issues with noise. Again, the noise portion will be covered in Section 4.5.2. Table 3 marks the major area of concerns for pass gate logic.

Pass Gate Circuit Family Sensitivities

The pass gate circuit family has the unique difference that a single NFET must try to pass both a logical 1 and a logical 0. The other circuit families in this chapter have the PFETs pass the logical 1 and the NFETs pass the logical 0. Using only an NFET pass gate, leads to a variation in performance depending on which logical value it is passing. When the NFET pass gate is passing a logical 1, the threshold voltage of the NFET is increased due to the body affected threshold. This will limit the output level of the pass gate to V_{DD}-V_{TB}, where V_{TB} is the body effected threshold. Therefore, a weak PFET "keeper" is desirable around the output invert just like the weak PFETs used in dynamic logic (the "keeper" is not shown in Figure 2.7).

L_{EFF} Variation	W_{EFF} Variation	V_T Variation	V_T, L_{EFF} Tracking	N/P V_T, L_{EFF} Tracking	Substrate Sensitivity	Leakage	Hot Carrier Wearout	Interconnect Resistance	Interconnect Capacitance	Logic Noise	Delay Noise	Supply Noise	Soft Errors (alpha,gamma)	Temperature	Supply Tolerance

TABLE 4-3 Pass Gate Circuit Family Sensitivities

The variations on delay as a function of process are shown in Figure 4.5. This figure shows the delay variations for the NFET pass gate passing both a logical 1 and a 0 with the gates of the NFET at V_{DD} and the signal approaching the source of the pass gate. This will be called either a "Pass 1" or a "Pass 0" case, respectively. This data assumed that 20% of the timing path was RC delay. If the last arriving signal is V_{DD} at the gate of the NFET pass gate and the data is a logical 1 passing through the pass gate, this will be called "Select 1". Finally, if the gate is the last to switch to V_{DD} and the NFET already has GND on the source, then this will be called "Select 0". The delay variations as a function of process for the "Select" cases will look nearly identical to the "Pass" cases. Those process parameters which affect V_T show twice the variation when passing a logical 1 than a logical 0. This is due to the body effect. Using the rule of the thumb that the body effected threshold voltage is twice as large as the standard threshold voltage, the variation in V_T will have nearly twice the variation in the body effected (V_{TB}). Therefore, V_T results in a much more sensitive process parameter (second only to L_{EFF}) when passing the logical 1. L_{EFF} again has the largest variation on performance of over 30%. This is larger than the ~20% variation for static circuits since the L_{EFF} will vary V_T (and V_{TB}) due to the short channel effect. W_{EFF} is again the smallest variable.

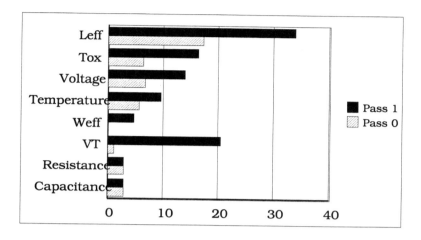

FIGURE 4.5 Delay variations of a pass gate circuits with process and environment.

Though not shown here, the delay of a pass gate circuit would again be a function of the weak PFET keeper. The goal of the keeper is to hold the level of the internal pass gate node high, to prevent any crossover or crowbar current. Similar rules of thumb apply to sizing the passgate device to the keeper device as in the dynamic circuits.

Driving the internal node to V_{DD} and reducing the crossover current in the inverter will also prevent the additional performance loss in the output inverter due to hot electron effects. The NFET pass gate is very sensitive to hot electron effect because it will spend a large amount of its time in the linear region when trying to pass a logical 1. Also, the current flow through the pass gate is bidirectional, since the source and drain change sides when passing a logical 1 versus a logical 0. Finally, since the hot electron effect results in a V_T change, and pass gates are sensitive to V_T changes, hot electron effect is more of concern.

One method of controlling the data dependency on the process variations is to make the NFET pass gates, full transmission gates with both an NFET and a PFET as in Figure 2.8 on page 69. Here the PFET is now responsible for passing the logical 1. Since the PFET will pass the logical 1 without degradation to the output signal, no weak PFET keeper is needed to hold the output high. The performance variations on the process parameters for selecting a logical 1 or selecting a logical 0 is shown in Figure 4.6. Again, L_{EFF} is the most prominent parameter, but the total variation is now about 20% for both passing or selecting a logical 1 or a logical 0.

Circuit Design Margin and Design Variability   **145**

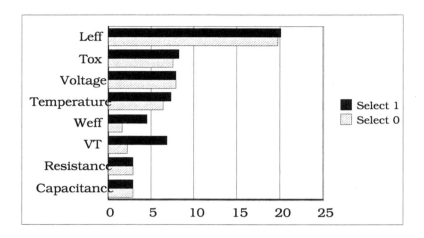

FIGURE 4.6 Delay variations of a transmission gate circuits with process and environment.

The performance difference due to variations in V_T is much reduced, but not as small as static gates. For layout considerations, the PFET is typically the same size as the NFET. This results in the variation due to V_T and W_{EFF} for selecting or passing a logical 1 If the PFET width is sized twice the width of the NFET, the W_{EFF} variation will drop to the same level as when selecting or passing a logical 0.

Since the pass gates act like a series resistance to the driving device and have a substantial amount of capacitance on the internal pass gate node, any process parameter that decreases the resistance of the pass gate advantageous.

The change in performance due to the variations in the wiring resistance and capacitance is nearly identical to other circuit families.

All pass gate circuit families mentioned in "Non-Clocked Pass-Gate Families" on page 65 have similar sensitivities to the process variations. Those with half-latches or with cross coupled outputs add the sensitivity of mistracking between the NFET and PFET for both performance and noise.

4.2.4 Differential Cascode Voltage Switch Logic (DCVS)

Table 4 shows the major areas of concern for differential cascode voltage switch logic circuits. (DCVS). A representative circuit for DCVS is shown in part (a) of Figure 2.4 on page 59. Like static circuits, the logic tree is comprised of NFET. The difference is that all inputs must have both the true and complement values available and the cross-coupled PFETs pulls and holds one of the outputs high. Since these PFETs are used to drive the output inverter and are not just weak PFET keepers, delay variations are susceptible to the tracking of the NFETs to the PFETs for both L_{EFF} and V_T. As will be seen in a Section Logic Noise has been remove from the list of sensitivities

DCVS Circuit Family Sensitivities

L_{EFF} Variation	W_{EFF} Variation	V_T Variation	V_T L_{EFF} Tracking	N/P V_T L_{EFF} Tracking	Substrate Sensitivity	Leakage	Hot Carrier Wearout	Interconnect Resistance	Interconnect Capacitance	Logic Noise	Delay Noise	Supply Noise	Soft Errors (alpha,gamma)	Temperature	Supply Tolerance

TABLE 4-4 DCVS Circuit Family Sensitivities

The DCVS circuit family shares the sensitivity of static circuitry to device transconductance; specifically, since the evaluate path commonly comprises NFET devices. The variation in delay is shown in Figure 4.7. The L_{EFF} is the major contributor to performance variation followed by the dependency upon voltage, temperature and oxide thickness. Generally, narrow channel effect and W_{EFF} variations is not a concern in DCVS families, as most of the unclocked styles must use wider devices to handle succeeding fanout load or to gain the switch point of the buffer load circuit.

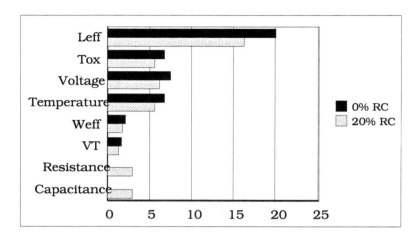

FIGURE 4.7 Delay variations of a Differential Cascode Voltage Switch circuits with process and environment.

The dependency of delay on V_T is similar to static logic and less of a concern than dynamic circuits. Therefore, the performance of the stacked devices found in the DCVS evaluate tree suffers from body effect. Static DCVS-like topologies are particularly vulnerable, as their logic evaluation trees must provide low resistance paths to ground in order to switch the load circuit; positive NFET source-substrate bias voltage increases threshold voltage and introduces additional delay. Changes in device substrate sensitivity will cause similar circuit response.

Evaluate loading is also critical. Capacitance on the tree slows transitions, as it must first be satisfied before the load book can be toggled. Process variations which increase capacitance, such as changes in source-drain capacitance, are reflected in performance. Enlightened layout to reduce junction area, position of the inputs on the tree based on arrival time, and tree device width tapering are techniques that are effective in minimizing delay.

DCVS topologies, such as DSL, which reduce the logic voltage swing to reduce power or delay are especially sensitive to parametric shifts which cause resistive voltage drop. As the signal voltage swing of a circuit is reduced, the difference in voltage between a stage's lowest possible "high" output and the lowest input still interpreted as a "high" from the following stage collapses, jeopardizing noise immunity. For the same reason supply tolerance also becomes an issue. As voltage swings are reduced for a given technology, the benefits of differential logic become significant. The dependence upon a precision reference bias voltage will are sensitive to threshold voltage control. A poorly controlled threshold voltage will result in a change in switchpoint and result in varying performance. Therefore, the effectiveness of differential feedback is strongly associated with the threshold voltage variation limits.

4.3 Design Induced Variation

Believe it or not, not all problems with circuits are related to process variations! Some designs have inherent variation within them, while a specific style will induce variations on performance and noise margins are a result of the specific implementation of a circuit. This section will deal with performance variations due to charge sharing, timing collisions, data dependent capacitance and die size. Each of these items will also effect the noise margins of a circuits. Noise margins will be discussed in Section 4.5, Noise on page 157.

4.3.1 Intermediate Charge and Charge Sharing

Charge stored on an intermediate node and then transferred to an output node can cause delay degradation or speedup. It can also cause noise margin degradation. For example, consider a two high stack of NFETs which are off and connected to an output node which at V_{DD}. If the bottom NFET is toggled on then off, the intermediate node is discharged to GND. Next the upper NFET in the stack is turned on. Some of the charge on the output node is moved to the intermediate node and the output node's voltage is dropped by a ratio of the intermediate node's capacitance to the output node's capacitance. This will cause some perturbation in the output state. Depending on the circuit family, this may cause a fail. Charge sharing is an issue for both noise margin and delay of a circuit.

A static NAND gate will have this configuration. Since static CMOS always has it's output driven, the history of the input signals have little effect on the output. A small dip may be seen on the output, but the complementary PFET will hold the node high. In this scenario, no effect would be seen upon delay. The induced noise should be small enough to not cause a fail. One should make sure that the amount of charge drawn off of the output node is not too large. Charge sharing in pass gates is not often seen, since two pass gates are rarely put into series. Also, DCVS circuits do not have a concern with charge sharing since the cross-coupled PFETs actively drive the output node.

Dynamic circuits will have the largest concern with charge sharing. If the output signal is truly dynamic, then the final voltage after the charge sharing event will be reduce by the ratio of the capacitance on the output node to the total capacitance of the internal and output node. Methods for reducing this charge sharing event are to add a weak PFET keeper device of adequate strength to hold the voltage high. A stronger keeper device will reduce the charge sharing voltage drop, but this will decrease performance as seen in Figure 4.4, "Delay vs. the weak PFET width.," on page 142. Another method for alleviating this problem it by precharging the intermediate nodes of the NFET stack. As the top devices turns on, there will be no charge pulled from the output node. This will add additional devices to the layout and increase the amount of capacitance on the internal stack devices because we added the precharge device to the this node and had to add a contact between the NFETs. This will slow down the speed of critical circuits. It will be more important than ever to put the last arriving signal at the top of the NFET stack because the body effect of the top device will be largest when the intermediate node is precharged. If the top NFET in the stack is the last to switch, its source will be at GND when the switching activity starts and giving it the best performance.

Even if one chooses not to use a precharge device for the intermediate node, speed may be sacrificed if the latest arriving signal is not on the top NFET. For example, a 2 input dynamic NAND may have the intermediate node precharged, but if the top device of the stack is first to switch, then the bottom device must remove more charge.The delay of a circuit can depend upon the recent switching of the input signals.

With zipper logic, one could imagine the same scenario with a predischarged node having a stack of PFETs attached to it. Two other non-elegant methods for reducing the amount of voltage drop due to charge sharing include: putting a very weak PFET device on the dynamic node which has its gate at GND. This method is not recommended in most applications since this will create additional DC current. The second method is to add more capacitance to the dynamic node, so that the capacitance ratio of the dynamic node to the charge sharing node is increased.

4.3.2 Timing Collisions

Performance and functionality can be impacted by the collision of two signals on the same circuit. Most notably, a requirement for a dynamic circuit is that the inputs must be ready before the precharge is released or the inputs must be monotonic upward during the evaluate portion of the cycle. If either of these scenarios are violated, then the precharge has been lost and a logical failure will occur. Other timing collisions include timing of a static circuit, multiple select lines selected on muxes and multiple output drivers.

Static circuits such as a NAND are timing dependent upon which input is the last to transition. If the top input to the NFET stack is the last to transition, it will be faster, than if both inputs switch at the same time. This timing collision is hard to account for. Most timing tools only account for one input switching at the same time.

Pass gate muxes typically have one select line at V_{DD}, so that the next node is always driven. This is called a one hot mux. When the desired select line switches from one pass gate to another, the opportunity exists to have two pass gates on at one time. If the data these two pass gates are passing is of opposite values, then the performance of the pass gate mux is severely degraded, because the two muxes are fighting each other rather than one pass gate pulling only static capacitance off the output. When they are fighting each other, the high current draw path is a short circuit from V_{DD} to GND through the PFET of one input driver, through two pass gates and to GND through the NFET of the other input driver. The internal node between the two pass gates will be determined by the voltage divider set up the this short circuit path. With this scenario, it has been seen to have a dependence upon the local interconnect within this path. If the resistance of the interconnect coming into the pass gate that is providing GND is large, then the internal node will be above $V_{DD}/2$. This will result in the output of the next inverter to remain low. Herein lies the example why pass gate circuits more susceptible to interconnect resistance than any other circuit style.

Multiple output drivers have the potential of having timing variations dependent upon the number of outputs that are connected together. For example, when an internal node of a one output (A) has an internal node that is also an output node for another timing path (B), then the timing of path A is dependent upon the variations of the output load on path B. This will result in widely varying delays to for path A (path B can be well characterized). This is typically not allowed as most timing tools do not have parameters of delay for path A related to the capacitance on path B.

4.3.3 Data Dependent Capacitance

The capacitance of a device varies as the FET changes from non-conducting to conducting. This capacitance is most noticeable on the gate of a device with different logical states on the source and drain. The data dependent capacitance can vary by more than an order of magnitude[4.4]. With the change in capacitance being most important on pass gates, this variation is very noticeable on the clocking network because most latches have the clocking signal attached to pass gates. (See Figure 5.3).

Since clocks inherently have a high fanout, the variation in the loading as the devices switch will have an appreciable variation of the rise/fall time and add to clock skew. The aspect that there is a large fan out also helps the clocking network because the probability of all the pass gates being in the worst case state for one cycle, then the best case situation for the next cycle is small and the loading reaches an average capacitance per gate width.

Logic circuits that fan out to different gate widths or to different polarity devices will have different loads and this will result in timing difference. With most timing rules generated from a static timing tool or from a SPICE simulation, the variation in capacitance is include in the timing description. Only the input loads are poorly defined by the timing descriptions.

4.3.4 Die Size Consideration

The size of the die for a particular chip sets the amount of process variation that one may see within chip. All process tracking parameters are set by the distance and dissimilar environment that two devices have with each other. Since L_{eff} is the largest driver of circuit performance variation, the across chip line variation (ACLV) is the most critical parameter. Larger chips have more across chip line variation within them and therefore have to accommodate larger variations in timing and clock skew because of the mistracking of the device length. The across chip variations can be reduced by keeping FETs that are in the same logic cone in close proximity. When this is not possible, one should keep the local environment the same (i.e. same physical direction and local polysilicon density).

For interconnects, the chip size also dictates the amount of variation with the thickness and width of a given wire and as well as the variation in the thickness of the interlayer dielectric. Some of these parameters are dictated by the radial component of the whole wafer. Chip size is also critical, since a single large chip would span a larger portion of a wafer's radius than a smaller chip. Whether the variations are due to a single chip's size or the location and size within a wafer, the interconnects variations results in a mistracking of the resistance and capacitance of every wire. In particular, the timing is more uncertain for global wires that travel a long way. This is particularly important for a clock distribution schemes that branches out from a central location without rebuffering at regional and local clock buffers. It is therefore less desirable to distribute the clock on a balanced tree or an H-tree with minimum width wires. This will lead to larger chips more easily have larger amounts of clock skew. To reduce the dependency of the clock tree upon the interconnect variations, buffers can be added to reduce the overall capacitance and to reduce the delay [4.5] .

Finally, a larger chip will have a larger percentage of the critical timing path dominated by the RC delay of the wires. This will result in a larger variation in the timing differences between two timing paths.

> **Rule of thumb:** The longest wire on a given chip is approximately the length of the longest side.

Larger chips also have the tendency to have greater thermal variations across them. This is a result of more function being put on a chip that is not used at the same time and the thermal resistance between to points will be larger the farther they are apart. Thermal variation result in timing variations and will be discussed in 4.4.2

4.4 Application Induced Variation

4.4.1 V_{DD} Tolerance Effects

The speed of any circuit can be increased by pushing the process to a faster condition (i.e. shortening L_{EFF}) or by raising V_{DD}. Have already shown the performance variations of different circuit styles as a function of voltage. Figure 4.8 shows the improvement in the frequency with an increase in V_{DD} for three different microprocessors in two different technologies. This microprocessor contains a combination of static and dynamic domino circuits in the critical path [4.6] . The base technologies were a 2.5V technology and a 1.8V technology. The increase in frequency of the fastest microprocessor is about 20MHz/100mV. This is an 8% variation in the performance for a 10% variation in power supply.

The rate of change for the frequency will not increase linearly as the applied voltage increases because the RC delay does not change as a function of voltage, the percentage of the critical path delay will include more RC delay at a higher voltage than it did at a lower voltage. Therefore, the frequency will start to flatten out as V_{DD} is increased. One method of decreasing the dependency upon RC delay is to add buffers to long wires. This will increase the speed of a critical path is done properly.

> **Rule of thumb:** Performance of a chip varies by 7-9% when the power supply is varied by 10%

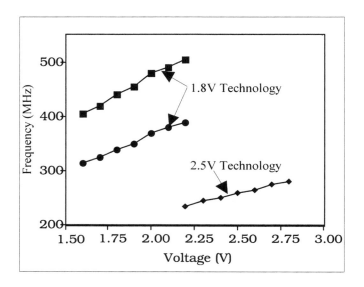

FIGURE 4.8 Frequency vs. voltage of three different microprocessors in two different technology.

Changing the external power supply varies the overall performance. Localized voltage drops on the internal power distribution system will create variations in delay across the chip. The power grid can induce both DC and AC voltage variations across the chip. Individual power spikes (valleys) in the localized power supply can be very large up to 20% of the power supply. Figure 4.9 shows the maximum voltage drop across a microprocessor. Here, the maximum drop is 7.5% of the power supply. The minimum drop is near 0%. The caches are located in the lower left and lower right corner of the chip and are identical structures, but they see a different localized supply voltage with a variations of about 3%. The cache in the lower left will be slower in performance. As expected, this variation is a function of time and depends upon the instruction being executed. If the chip is in a low power mode, the voltage difference across the chip is very nearly 0%. Varying circuit styles will result in larger fluctuations. For example, if dynamic circuits are chosen as the predominant style, then the clocking distribution will consume most of the power. Some microprocessor have 40% of their power consumed within the clocks [4.7] . The distribution of power on an active circuits changes the tolerance of delay and noise margin on all circuits. A well built power distribution system may still have a steady state voltage drop of up to 10% of the supply voltage when the chip is active.

> **Rule of thumb:** A typical circuit should be able to handle internal voltage drops of 5-10% of the V_{DD}.

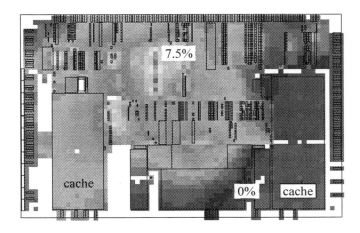

FIGURE 4.9 Voltage drop across the power distribution of a microprocessor.

Decoupling capacitance can be added to the power supply to reduce the noise on the supply distribution. However, decoupling capacitance must be electrically close to the source of the noise to be able to suppress the noise. On module decoupling capacitance have little effect on the internal localized supply noise since the inductance and the series resistance will shield and delay the charge that the decoupling capacitance could provide. On chip decoupling capacitance will have the largest effect on suppressing supply noise. The largest source of internal decoupling capacitance is the capacitance that is connected through the active devices. This capacitance is limited by the series resistance provided by the FETs. The second largest source of internal capacitance is the NWELL to substrate diffusion capacitance. This diffusion capacitance has a large area, but does not hold much charge in a given area. Also, the series resistance of the NWELL severely limits the ability of the charge to move back to the supply rails in a timely manner. The best on chip capacitance is a the thin oxide capacitance of a FET. Directly connected FETs provide a large amount of decoupling capacitance. One problem with the FET as a capacitor is that it too takes a large amount of area to produce much capacitance. Defects within the oxide can result in reduced yield if the defect shorts the power supply rails. This can be eliminated by adding a switch in series to turn off the DC current on shorted oxides. This internal networks that is in series with the decoupling capacitance will again limit its the usefulness.

Inductance in the power distribution system can also lead to additional noise in the power supply. The inductance can come from the packaging and external interconnects. This inductance can create an oscillatory effect on the power supplies. Methods for reducing the inductance is listed in [4.12] .

4.4.2 Temperature Tolerance Effects

Since the mobility of electrons and holes decreases with the increase in temperature and the current through an FET will decrease, the circuits that operate at a higher temperature will be slower than circuits operating at cooler temperatures. Also, a reduction in temperature will reduce the resistance of the metal interconnects. Therefore, a reduced temperature can be used to one's advantage by not allowing any portion of a chip to heat above a certain temperature by limiting the power consumption, improving the thermal resistance of the material, package or environment, or by decreasing the external temperature of the environment. This reduction in the maximum temperature will result in a higher performance chip. A 20 degree celsius decrease in the maximum operating temperature will improve the rated frequency of a chip by nearly 3%. Many systems take advantage of a reduced temperature to gain additional performance.

> **Rule of thumb:** Circuit performance varies about 1.5% per 10 degree Celsius

Aside from the maximum operating temperature, the localized temperature gradient across a chip is not uniform. It can vary significantly depending on the operation that the chip is performing, the size of the chip and the thermal resistance across the chip. In a low power mode (such as sleep mode), the temperature is fairly uniform because no instructions are being executed and little power is being dissipated, while in full operating mode where an intensive set of instructions such as a floating point instruction, the variation across a chip may be as high at 15 degrees C [4.8]. Stating the above rule of thumb, this could lead to a skew in timing between two paths of nearly 2.3%. The temperature gradient will also result in additional skew in the clocking distribution, most notably due to the variation in the change of resistance of the interconnect.

4.5 Noise

Noise within CMOS digital circuits is more prevalent today than ever. The reasons for this is mostly due to the continued shrinking of the physical chip dimensions and interconnects within the chip. As the separation distance between interconnects decrease, the amount of capacitance to another active interconnect increases. Therefore, any transition on a given wire will induce some noise onto any other wires in the surrounding area. This is called capacitive coupling or crosstalk. Crosstalk effects on the specific circuit styles will be the main focus of this section. Additional noise can be related to the sagging and recovering of the power supplies. Since the power supply network is a large RLC circuit, it is possible to set up ringing on both the V_{DD} and GND power supplies. Noise associated with transmission lines and their discontinuances will be discussed in Chapter 6.

4.5.1 Capacitive Coupling (Crosstalk)

Any two metal wires that lie next to each other are electrically isolated by the dielectric that lies between them, but are electrically connected by the capacitance across this dielectric. The amount of current that is exchanged by the two wires is dependent upon $i=CdV/dt$, where C is the capacitance between the wires and dV/dt is the rate of change of the voltage between a wire and its neighbor. The capacitance is a function of the width of the wire, the linear space between the lines, the thickness of the wire, the separation distance to any other conductor such as wiring levels above and below, the type of dielectric and the distance of common run length shared by the two wires. As technologies continue to scale to smaller dimensions and circuits continue to get faster, the amount of crosstalk between neighboring wires is increasing since the edge rate of signals is faster and the ratio of the coupling capacitance to a neighboring wire to the total wire capacitance on a given wire is higher. Voltage pulses placed upon one wire (victim) when another wire (aggressor) switches is the essence of crosstalk noise.

Figure 4.10 shows the height of the noise pulses generated on a 1mm victim wire for various percentages of coupling capacitance to total capacitance. In this scenario, the victim wire is held to ground by an NFET with various widths while the aggressor wire had a rise time of 100ps. The source and sink for the victim and aggressor wires were in close proximity to each other. The noise for this data was measured at the far end of the 1mm wire. The noise is slightly reduced at the near end of the 1mm wire since the victim wire's driving NFET is not resistively shielded from the noise pulse.

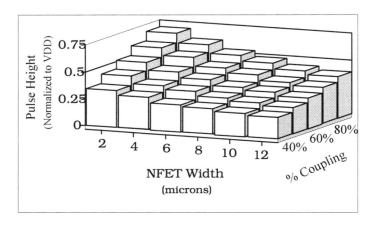

FIGURE 4.10 Noise pulse amplitude as a function of the NFET width on the victim wire and as a function of the percentage of coupling from the aggressor wire to the victim wire.

As shown, the noise pulse on the victim wire can vary from 25% to 75% of the power supply. If the victim net is not held by a very large NFET device, the amplitude of the noise is large. Also, the amplitude of the noise is larger when a higher percentage of the total capacitance is dominated by coupling capacitance. To get higher than 50% of the power supply, it would require that the victim wire be held by an NFET width of less than 8 microns or the coupling percentage to be less than 60%. With 0.25υm technologies, the coupling percentage is in the range of 60-70% for tightly nested wires.

Different wiring configurations will result in different noise amplitudes and widths. One can make this noise pulse larger by putting the driver of the victim wire farther away. For example, if the victim net was 2mm long and the aggressor net overlapped the victim net for 1mm, but started in the middle and extended to the far end, this would shield the victim wire's driver even more. Another configuration is if the aggressor wire's signal was traveling in the opposite directions. Therefore, the noise event would start very near the sink on the victim wire. Again, this would result in the victim wire's driver being resistively shielded from the noise pulse at the far end of its wire. If one makes the wires longer and they overlap more, the noise pulse can be taller or it may just become wider due to the dispersion of a pulse travelling down a resistive line. This would result in a slightly shorter pulse that is wider. We will see in subsequent sections that this is not necessarily any better.

It is possible to have a similar amplitude on a negative going pulse on a victim line if the victim line is held to GND and the aggressor wires transitions from V_{DD} to GND. The same coupling applies and amplitude results, but the PFET on the driving net will start to turn on as the victim wire drops below its threshold voltage and if the victim net drops below a diode forward bias voltage, then the junction of the drain of the NFET will forward bias and conduct to the substrate. This will limit the height of a noise pulse near a driver.

It is possible to have an infinite number of possible noise events between wires. As many different aggressor wires are possible switching either in an upward or downward transition.

> **Rule of thumb:** One only needs to account for the 3-4 longest aggressor wires for any particular victim wire.

Methods for reducing the crosstalk from a process point of view include making the wires thinner, not allowing tight minimum spaces, low epsilon dielectrics. From a design perspective one can space the wires farther apart, keep the wires as short as possible, increase the driving strength of the victim lines, make sure the direction of transitions of the neighboring wires are in a direction that do not cause logic noise.

We will not discuss noise pulses that arrive at the driving circuit for a victim wire. These noise pulses are easily quenched by that driving circuit and are rarely of concern for CMOS circuits.

4.5.2 Delay Noise

When induced noise from the aggressor wire either speeds up a victim wire or slows it down, this is called delay noise. It is synchronous in nature since both the victim and aggressor net are switching at the same time. If the victim net is on an upward transition, and the aggressor net is also on an upward transition at the same time, then the dV/dt is very small and the driver of the victim net will have less effective capacitance to charge and this will speed up the driving circuits delay and reduce the RC delay of the victim. This may induce a race condition if the decreased delay causes the signal on the victim net to arrive too soon at its destination circuits. For example, consider a data path that is sped up due to coupling and arrives at a latch, before the latch is closed to prevent the current correct data that is within the latch to be maintained. This is a failed race condition where same direction switching can cause chip failures. The counterpart to this example is when the victim net and the aggressor net are switching in the opposite directions. Here the voltage change across the coupling capacitance is now 2dV/dt and results in a larger effective capacitance for the driving circuit to overcome. This added delay could result in a signal being propagated too slow and failing to arrive at a latch in time to overwrite the latch with the correct data. Performance limited paths that consist of a large amount of RC delay must accommodate the added delay due to switching neighbors to accurately predict a chips final cycle time. Delay can be a function of the inductance of a line. Inductance will not be considered in this chapter [4.9] .

4.5.3 Logic Noise

Speed variations from delay noise is not the only issue with crosstalk, logic noise and functionality are of concern too. Logic noise is defined as an asynchronous noise pulse from a neighboring net onto victim net which is quiet which causes a circuit fail. The forward propagation of this logic noise onward into a cone of logic is of utmost concern because chip failures will result if an incorrect value is captured within a latch. We will now consider the same CMOS circuit family (static, dynamic domino, pass gate, and DCVS) and the effect of noise arriving at their inputs. We will look at the passing and failing regions for each circuit family as a function of the input noise pulse height and the input noise width. This will result in a noise schmoo with the passing region being below the curves for positive going pulse on a wire that is at GND. Only pass gate circuits will be concerned with negative going pulses on a wire that is at GND. Due to the symmetry of the problem we will not investigate the negative going pulses on wires that are at V_{DD} or positive going pulses on wires that are at V_{DD}.

Static CMOS Circuits

Static circuits will eliminate any pulse from being transmitted if the height is too small or the width is too narrow. When this occurs, the noise margin of the receiving circuit is adequate to prevent the pulse from being amplified and propagated. If the noise pulse is larger than the noise margin, the pulse will be amplified by the static circuit and the next static circuit will now be the determining factor on whether the pulse is propagated further or not. Noise is always recoverable in static circuits provided enough time is allotted for the propagated noise pulse to be corrected.

Figure 4.11 shows the noise schmoo for a static inverter with a PFET to NFET width ratio of 2:1. The open circles is for a nominal process and environment. The minimum and maximum curves are for the whole process window, with an 8% supply tolerance and a 50 degree Celsius temperature tolerance. Again, the passing region is below the curves. A fail for an inverting circuit is defined when the output has decreased by a larger amount than the input pulse height. This means that the inverting circuit has provided gain to the noise pulse.

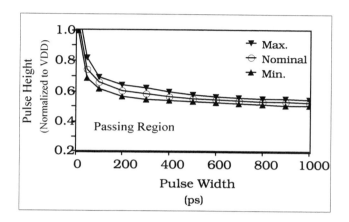

FIGURE 4.11 Noise schmoo for static CMOS circuits including process and environment variations.

For very wide pulses that approach DC, the maximum height for an input pulse is approaching $V_{DD}/2$. This equal to the switch point for a 2:1 ratio of an inverter. If a different inverter ratio is used, the noise schmoo would look very much the same, but the DC value would approach the respective switch point. As the noise pulse gets shorter, the pulse height may get larger and the inverter will not pass the noise pulse. The noise pulse width will be nearly equal to the rise/fall time of the aggressor wire, because that is when the charge is move across the coupling capacitor. The minimum and maximum variation are not very strong for static circuits. The largest contribution to the shift in the noise schmoo is the mistracking between the NFET and the PFET of the inverter. As the NFET gets stronger than the PFET, the pulse height that would pass through an inverter will be smaller.

As the technologies get faster and faster, the noise schmoo has the same basic shape. The difference will be seen in the passing regions upward curve will occur at narrower and narrower pulse widths. This does not dictate that the noise will be worse, because the pulse width of most noise pulses will be narrower at the same time due to the sharper rise and fall times of all circuits.

A victim wire that is held at GND in static logic could have a negative going pulse put upon it by an aggressor net that is transitioning from V_{DD} to GND. The pulses amplitude is negative and will not be transmitted by static logic. Likewise, signals held at V_{DD} and have positive noise pulses above V_{DD} will not be transmitted through static circuits.

Logic Noise in Dynamic Domino Circuits

For noise to propagate through a dynamic domino circuit, it must start to discharge the intermediate dynamic node and then pass through the output inverter. Figure 4.12 shows the noise schmoo for dynamic domino circuits across process variations.

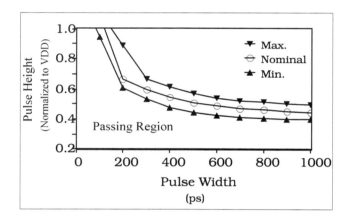

FIGURE 4.12 Noise schmoo for dynamic CMOS circuits including process and environment variations.

Dynamic circuits are more sensitive to noise because one of the two switching thresholds is dominated by the threshold of the NFET pulldowns as mentioned in Section 4.2.2. Therefore, an incoming noise pulse will start to discharge the dynamic node to the domino circuit much more easily. The process spread on the noise margins are greater than they were for static circuits due to the increased sensitivities of this logic family. The DC value of the dynamic circuit is lower than static circuits. As a design trade-off, one may improve the noise margins shown in this schmoo by adjusting the strength of the weak feedback device and slowing down the speed of the domino circuit. Another way of adjusting the noise schmoo upward is to make the switch point of the output inverter lower by decreasing the PFET to NFET width ratio. This again will slow down the dynamic circuit. Dynamic circuits will not propagate pulses that have an amplitude below GND.

If the circuit family was a zipper domino or another dynamic family that did not have the static inverter, then the noise schmoo would be much lower and the process variations would be much greater. The inverter and half latch improve the noise margin significantly [4.11] .

Logic Noise in Pass Gate Circuit

Pass gate circuits can have noise approach one of its two different inputs, the select line on the gate of the NFET or the data input on the source side of the FET. Considering an NFET only pass gate with GND on the gate, V_{DD} on the drain and GND on the source. If a positive noise is applied to the gate, this will be called "Select 0" since the noise is on the select line and a logical 0 may pass through. If positive noise is on the gate when the NFET is off and the data input is at V_{DD}, while the internal node is at GND, this will be call "Select 1", since a noise pulse on the select line may let a logical 1 pass through. Finally, if a negative noise pulse is on the data while it is at GND, he gate is at GND and the internal node of the pass gate is at V_{DD}, this will be called "Pass 0" since a logical 0 may pass through the pass gate. These three cases are shown in Figure 4.13. The "Pass 1" case is not a considered since a positive pulse on a source of a pass gate held at V_{DD} will not create any noise propagation.

When a positive going pulse is applied to the gate, the height and width of the pulse is critical to determining if a fail will occur as well as the logical values on the source and drain of the NFET. In the both the Select 1 and Select 0 cases, the noise pulse on the gate must has enough amplitude rise above V_T and turn the NFET on, then the I_{DS} of the NFET may dissipate enough charge from the drain node to cause a failing value to pass through the NFET.

Notice that the Select 0 case is very similar to the static and dynamic circuits. The DC value is only slightly smaller than either case and the process sensitivities (not shown) are very similar to the dynamic circuit family. The circuit sensitivities are closely related to the NFET to PFET device tracking and the threshold voltage of the NFET pass gates. To increase the passing region, the same trade-offs apply. One may strengthen the weak feedback PFET and/or lower the output inverters switchpoint by widening the NFET of the inverter. Both of these measures will decrease the performance of the pass gate circuits.

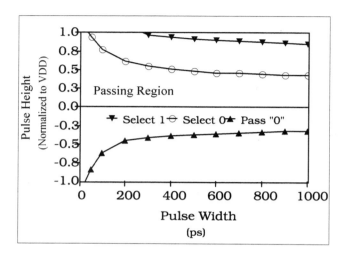

FIGURE 4.13 Noise schmoo for pass gate circuits for noise on the select line and on the data input.

The Select 1 case is significantly different than any circuit family we have seen so far. To pass a logical 1 through the pass gate with noise on the gate requires a very wide pulse with a large amplitude. This reason that this is so robust is the same reason that this transitions is so slow to in the functional mode. With an NFET pass gate, the body effected threshold voltage greatly reduces the ability of the NFET to pass a logical 1. Therefore, it is difficult to fail by passing a logical 1. As with the performance variation, the Select 1 case has the most process sensitivities of any noise issue discussed. This again is due to the sensitivity of the body effected threshold voltage to the control of the threshold voltage of the NFET. Few design options exist to move this noise schmoo curve for Select 1, but it is not usually a concern unless the threshold voltages of the pass gate NFET get very small.

The Pass 0 case is a common failure mode. Notice in Figure 4.13 it requires a negative amplitude. When the magnitude of the negative noise pulse on the data input is greater than the threshold voltage, the NFET exits the subthreshold region and starts to conduct current. The charge is drawn off the internal node and the a logical 0 is passed through the pass gate. This fail is very common in scan latches that have very long interconnects into the input of the latches. The same trade-offs exist to move the magnitude of the noise schmoo curve higher as with the Select 0 case. A unique fix for this noise pulse is to put an inverter in front of the data input. Since an inverter will not pass a negative going noise pulse, and a pass gate will not pass a positive going noise pulse on the data input, the combination of these two circuits will eliminate most noise concerns.

Some noise pulses will not have any effect on the pass gate circuits. If a negative going pulse is applied to the gate of an off NFET, then the signal is not transmitted except by the Miller capacitance between the gate and the drain and this is small enough not to cause any the noise to propagate. A positive going pulse on the data input when the data input is at GND or V_{DD} will not pass through an off pass gate.

These ideas are directly extendable to full complementary transmission gates. With the addition of a PFET in parallel with the NFET for the transmission gate, a new failing mechanism is possible. When the data input is at V_{DD} and a positive noise pulse arrives, then the PFET can turn on and pass a logical 1 through the transmission gate. This could be called a Pass 1 case and would look very similar to the Pass 0 noise schmoo, except the pulse height would be positive.

Finally, these noise schmoos where for noise on one input. If the pass gate circuit is a wide mux structure, the noise schmoo may be lowered if noise is applied to more than one pass gate at one time.

Logic Noise on Differential Cascode Voltage Switch Logic (DCVS)

DCVS is the most noise immune circuit family discussed in this book. Two separate fail mechanisms will be considered. Consider Figure 2.4 on page 59, if a positive noise pulse is present on the input A and is propagated with equal magnitude to the output signal NAND or $\overline{Q}$, this will be called A-NAND in Figure 4.14. The second case is the same as A-NAND, but adds a second simultaneous negative noise pulse to the input $\overline{A}$. A fail is defined when both NAND and AND see noise of equal magnitudes as both of the inputs. This case is called "Double" in Figure 4.14.

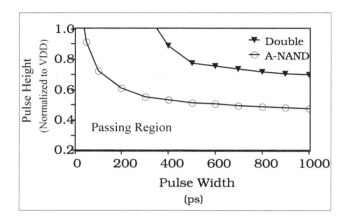

FIGURE 4.14 Noise schmoo for DCVS circuits including two fail mechanisms.

The noise schmoo for the A-NAND case is reminiscent of the other circuit styles. With typical dimensions, the DC failing region start near $V_{DD}/2$. A single sided noise event such as A-NAND will not result in a permanent fail like a dynamic domino noise fail. Given enough time, the cross coupled PFET holding output NAND high will again restore to the correct output states because the complement signal will hold the cross coupled PFETs in the proper state and the noise will not be permanent. Any noise that does propagate through onto the output must be fighting a PFET which is providing very local to the current circuit. This noise pulse should be quenched by the subsequent circuit. While the output NAND is being pulled low during the noise event, the PFET holding the output AND is being turned on. Since the noise pulse is not usually very strong and the cross coupled PFETs are much weaker than the pull-down NFETs, little perturbation will be seen on the other output AND.

Noise propagation to both outputs is possible if noise is present on both the A and $\overline{A}$ inputs. For this fail, the noise must be a positive going pulse on the input that is at GND and a negative going pulse on the input that is at V_{DD}. This is not likely to happen at the same time. If this did happen, the noise schmoo would look like the curve labeled Double. The DC value for the input noise would have to be 70% of supply rails. Any pulse width less than 300ps wide would not cause a fail. In 0.25µm technologies, most rise and fall times are near 100ps. If a noise pulse is created that is wider than 300ps it is unlikely to have an amplitude needed to create a fail. Therefore, it is very unlikely for a noise event to be propagated to both outputs of a DCVS circuit.

Just like static and dynamic circuits, DCVS logic gates will not propagate pulses that have an amplitude below GND or pulses that rise above V_{DD}.

4.5.4 Transmission Lines

Interconnects between two CMOS circuits can present different loads to the driving circuit depending upon the environment that they are in. An interconnect in today's technologies were the wire is about as tall as the interconnect is wide for a minimum pitch line results in the coupling capacitance between two wires to be about 65% of the total capacitance on a wire. The resistance of the interconnect can be large (several hundred ohms) for a very long wire.

If one treats the interconnect as a simple lumped capacitor and resistor, the driving circuit will see too large of a load and the rise delay of the signal at the output of the driver will be too slow. This is not realistic as the capacitance and resistance is distributed over the length of the interconnect[4.13]. The capacitance at the far end of the interconnect is shielded from the driver by the series resistance. Therefore, the output of the driver will rise quickly. The signal will propagate down the transmission line and dispersion causing the rise time at the far end of the transmission line to increase. The environment of the interconnect line will change the rise/fall characteristics of the propagating signal.

> **Rule of thumb:** Interconnects must be treated as distributed lines for correct loading of driving circuits.

4.5.5 Simultaneous Switching

Simultaneous switching of neighboring circuits cause simultaneous need for current to be supplied by the power distribution grid. This can cause V_{DD} power rail to droop during the switching events. This is most prevalent with I/Os. Usually the I/O's power rails are kept separate from the internal chip power distribution system for two reasons. First, the I/Os may have a different output voltage than the internal voltage of the rest of the chip. For example, microprocessors in current technologies have internal voltage of 1.8 V and the I/Os must interface with a system board that is at 3.3V. Secondly, the added noised that I/Os generate on the I/O's power supplies is kept off of the internal power distribution system if these rails are kept separate. This is good for the internal circuits since there power supply noise is now reduced. However, this is a disadvantage for the I/O's power supply since there distribution system is now smaller since it is kept separate from the rest of the chip and there is much less decoupling capacitance for it to draw upon. Another problem with I/O power supplies is that when simultaneous switching occurs, the demand upon the distribution system are quite large since each I/O need to drive the very large capacitance of the system. The local voltage of an I/O can droop as much as 25% when all of the I/Os are switching at one time. To reduced the amount of droop on the I/O power supply, one typically will limit the logic from switching all the I/Os at once, or will add external decoupling capacitors to the module. Additional discussion of I/Os will be given in Chapter 6.

Clocked circuit styles like dynamic domino circuits inherently have more noise on the power supplies because the clocks switch a larger amount of capacitance every cycle. This added noise subtracts from the noise margins allowed within a chip when designing a circuit style with a noisy supply. The power supply and distribution can have a damped oscillation at a resonant frequency of the RLC network. Additional V_{DD} droop and GND bounce will take away from the overall noise margins of circuits. Some of the inductance is driven by the packaging and off chip interconnects.

4.5.6 Soft Errors

Soft Errors are errors which can be removed by rewriting the area of lost data [4.15] . Unlike hard errors, which are often manufacturing defects, soft errors can occur at any time and are from radiation particles such as alpha or gamma particles. As discussed in Section 1.3.4, these incident particle will produce additional electron-hole pairs and will add or deplete charge from a local node. The amount of charge needed to retain a given logical value on the node that experiences the additional electrons or holes is called Q_{CRIT}. A node may be disturbed if the Q_{CRIT} is too small from too little capacitance or the active devices attached to the node are too weak.

Increasing Q_{CRIT} requires more area. The most common method of adding additional capacitance to a node is to increase the size of the source and drain junctions. If one chooses to strengthen the holding devices, this will usually add additional junction capacitance too. Adding junction capacitance means that one also has increased the junction area and hence increased the probability of capturing an alpha particle or a gamma particle. If space permits, one can always design a circuit that is immune to normal levels of radiation. Some applications, such as electronics in satellites, require radiation hardened circuits and demand that the larger memory cells and latches be use.

Each design must design for an acceptable Soft Error Rate (SER) that it can tolerate. Like crosstalk, the induce charge on a given node is more perilous to certain circuits. It is most commonly a problem in memory cells, then dynamic circuits and finally latches. In each of these circuits, the devices that are maintaining a logical 1 are weak feedback PFETs. Also, each of these circuits cannot recover from a soft error event on its own. A static circuit or a DCVS circuit, would recover from any incident particle. Also, static and DCVS will have more capacitance and larger PFETs driving the output nodes and the extra charge will not create as large of voltage variations.

Memory cells are the most susceptible to soft errors. Since there are so many memory cells within a chip, one must keep them small in size. This size constraint leads one to reduce the amount of charge stored on all internal nets within a memory cell and reduce the size of the holding devices. With large caches and many different levels of caches on chip, probability of an alpha particle intersecting a memory cell is larger than the probability of a memory cell intersecting any other circuit.

> **Rule of thumb:** Memory accounts for 25-45% of the area of a microprocessor with L1 caches on chip, while latches account for < 10% of the area of a microprocessor. If a microprocessor has both and L1 and an L2 cache on chip, the percentage of area used for the caches may rise to 75%.

Circuit methods for detecting SER include parity and Error Correction Coding (ECC) [4.14] . With parity, one adds an extra bit to every byte to force the number of ones within the 9-bit segment to be even or odd. This is called even or odd parity, respectively. ECCC adds detection and correction capabilities. Typical ECCC implementations are added such that two soft errors can detected and one soft error can be corrected. This adds addition bits that one must have within the memory. Also, this adds overhead to the logic of a microprocessor to be able to detect and/or correct the failing bit(s). This overhead can come in the form of reduced performance due to longer cycle times or deeper pipelines.

As memory within a chip becomes larger, the SER must be considered and most applications require that correction and detection be part of the memory scheme to produce the appropriate fail rates.

4.6 Design Margin Budgeting

Budgeting for design margin is required, so that one does not lose functionality over a process range or under a given set of application conditions. However, as a designer, one would rather not be too conservative and design out all of the possible speed within a design to provide more margin. A trade-off must be provided between the margins for noise, SER and circuit performance.

Performance variations will total in the across chip process variations, the on chip supply voltage drop and the on chip temperature differences for fast path analysis while the slow path or critical path performance is dictated by the global process and application ranges.

For the minimum performance capabilities, the slowest set of circuits will be the at the slowest process corner, the lowest supply voltage with the largest on chip voltage drop and at the highest temperature. Each of the parameters are assumed to match across the chip and provide global performance variations. The maximum total delay is given below, where t_{cycle} is the cycle time, t_{latch} is the launch delay from the latch, t_{setup} is the setup time into the next latch, t_i is the delay of each circuit between the latches and RC_i is the interconnect delay between circuits. The interconnect delay

$$t_{cycle, max} = t_{latch,max} + t_{setup,max} + \sum t_{i, max} + \sum RC_{i, max} \qquad (4.1)$$

should account for delay noise from cross talk as well as the resistance and capacitance of the wire.

Localized process variations do not affect the overall performance much since performance limiting paths have a large number of gates per cycle and the variation is averaged out over each stage. As cycle times are pushed shorter, the number of gates per cycle is decreasing and the localized variation is more critical.

The paths were localized performance differences are considered to be of utmost importance are very short paths between latches. Even though clocks are designed to have non-overlapped phases, the process variations on RC and clock generators can result in clock skew where the phases will overlap. In a master-slave latch configuration, this may result in data passing too quickly from slave portion of one latch through the master portion of the next latch. The "race" condition is between the clocking and the data signals. In the equation 4.2, the maximum clock skew is the sum of the maximum difference in the interconnect delay and the delay through clock generator and local clock buffers. The interconnect delay varies due to thermal differences, resistance and capacitance differences of two different branches of the clock tree and is denoted by ΔRC_{max}. The difference in the generator and buffer delays are due to within chip thermal differences, within chip process variations, and within chip power supply variations. This is denoted by $\Delta t_{clock,max}$. determined by the across chip process, supply voltage and temperature variations.

$$skew_{max} = \sum \Delta RC_{max} + \sum \Delta t_{clock,\,max} \qquad (4.2)$$

A fast path condition that fails is listed in equation 4.3. Here the $t_{hold,max}$ is the hold time for the latch. The sum of the minimum interconnect delay and minimum gate delays for the fast timing path must be more than the sum of the maximum clock skew and the maximum hold time for the latch. Again, the minimums and maximums are

$$skew_{max} + t_{hold,\,max} < \sum RC_{i,\,min} + \sum t_{i,\,min} \qquad (4.3)$$

within chip variations, not total process window variations. The minimum and maximum variation within the chip will vary across the process window. Since clock skew is dominated by RC variations and the fast path delay is dominated gate delay, it is best to use the within chip variations at the fastest process, highest voltage and lowest temperature to meet the required condition of equation 4.3.

Noise margin budgeting is similar to fast path analysis in that the noise margins are degraded by variations of the parameters within the chip. Equation 4.4 shows the total noise margin allotment. Where $V_{crosstalk}$ is the worst case noise due to crosstalk

$$V_{noise,\,total} = V_{crosstalk} - \Delta V_{supply} \qquad (4.4)$$

at the worst process corner, ΔV_{supply} is the worst case droop in V_{DD} plus the rise in GND for both chip power distribution and package power distribution.

4.7 Summary

Process and application conditions vary and effect performance and noise margins. Total process variations affect overall performance, while localized process variations effect noise margins and fast path timings. In a similar manner, global power supply voltage and temperature will effect performance, while the localized variations have a bigger effect on noise margins, clock skew and hence, fast paths.

Static CMOS circuits are the most robust family and are the least sensitive to process variations. The switch point of static CMOS circuits is most heavily dominated by mistracking of the NFET to the PFET. Crosstalk does not create logic noise given enough time, but will create delay noise.

Dynamic domino circuits are the fastest circuit family. The process variations provide more variation in performance and noise margins. Noise margins can be improved by adding a weak feedback PFET device around the output at the sacrifice of speed. Dynamic circuits consume more power and will contribute to a larger variation in the droop of the power supply if not designed correctly.

Pass gate circuits are the most sensitive to process variations. With the body effect producing a larger V_{TB}, the circuit is very sensitive to V_T variations. Performance varies widely with L_{EFF}, since L_{EFF} also varies the threshold voltage. Noise is prevalent on both the gates and the sources of the pass gates. It is very hard to create noise that will pass a logical 1, but only pass gate circuits will fail with a negative going pulse when applied to a source that is already at GND. Full transmission gates will improve the sensitivity to the threshold voltage and will add another logic noise mechanism.

Differential cascode voltage switch circuits are the most noise free circuits of the four explored in this chapter. Like static circuits, given enough time, the logic noise will correct itself and should only be considered delay noise.

REFERENCES

[4.1]   Qian, J, S. Puellel, and L. T. Pillage, "Modeling the Effective Capacitance of RC Interconnects," *IEEE Trans. Computer-Aided Design*, May 1993.

[4.2]   Hodges, D. and H. Jackson, *Analysis and Design of Digital Integrated Circuits*, New York, McGraw-Hill, 1988, pp.85-88.

[4.3]   Sedra, A. and K. C. Smith, *Microelectronic Circuits*, New York, Holt, Rinehart and Winston, 1987, pg.52.

[4.4]   Fair, H and D. Bailey, "Clocking Design and Analysis for a 600MHz Alpha Microprocessor," *1998 IEEE ISSCC Digest of Technical Papers*, Feb.5-7, 1998, pp.398-399.

[4.5]   Pullela, S., N. Menezes, and L. Pileggi, "Post-Processing of Clock Trees via Wiresizing and Buffering for Robust Design," *IEEE Transactions on Computer-Aided Design of Integrated Circuits and Systems*, Vol, 15, No.6, June 1996, pp. 691-701.

[4.6]   Rohrer, N., et al. "A 480 MHz RISC Microprocessor in a $0.12\mu m$ L_{EFF} CMOS Technology with Copper Interconnects," *IEEE ISSCC 1998 Digest of Technical Papers*, Feb. 5-7, 1998, pp. 240-241.

[4.7]   Gronowski, P., et al., "High-Performance Microprocessor Design," *IEEE Journal of Solid-State Circuits*, Vol.33, No.5, May 1998, pp. 676-686.

[4.8]   Geannopoulos, G. and K. Dai, "An Adaptive Digital Deskewing Circuit for Clock Distribution," *IEEE ISSCC Slide Supplement to the 1998 Digest of Technical Papers*, Feb. 5-7, 1998, pp. 326-327.

[4.9]   Shoji, M., *High-Speed Digital Circuits*, Reading, MA, Addison-Wesley Publishing Co., 1996, pp 237-241.

[4.10] Shoji, M., *Theory of CMOS Digital Circuits and Circuit Failures*, Princeton, NJ, Princeton University Press, 1992, pg. 196.

[4.11] Larsson, P. and C. Svensson, "Noise in Dynamic CMOS Circuits", *IEEE Journal of Solid-State Circuits*, Vol. 29, No. 6, June 1994, pp.655-662.

[4.12] Buchanan, J., *Signal and Power Integrity in Digital Systems*, New York, McGraw-Hill, 1996, pp.89-91.

[4.13]   Elmore, W. C., "The Transient Response of Damped Linear Networks with Particular Regard to Wide-Band Amplifiers," *Journal of Applied Physics*, Vol. 19, No. 1,Jan. 1948, pp.53-63.

[4.14]   Fifield, J. A. and C. H. Stapper, "High Speed on Chip ECC for Synergistic Fault-Tolerant Memory Chips," *IEEE Journal of Solid-State Circuits,* Vol. 26, No.10, Oct. 1991.

[4.15]   Ziegler, J. F., et al., "IBM experiments in soft fails in computer electronics" (1978-1994), *IBM Journal of Research and Development*, Vol.40, No.1, Jan. 1996, pp.3-18.

CHAPTER 5

Latching Strategies

5.1 Introduction

Latching refers to the system and usage guidelines that govern a given chip design with respect to storage elements, which are circuits that store digital signal states. Latching is one of the most critical areas in today's chip designs. Since nearly all microprocessors and computer systems are pipelined, storage elements are required to form the partitions for nearly every pipeline stage on the chip. Because they bound the logic, latches are also crucial elements for several components of the design methodology for the chip. Latches are the starting and ending points for most all timing and test analysis performed on a chip design. The testing and bring-up of early hardware samples of a new design depend more upon the successful operation of the latches than most other circuits on a chip. Consequently, without a good solid latching structure and strategy, the entire chip and system design is negatively impacted. This chapter will focus on types of storage elements and their design, how they fit into an overall chip system, and the types of errors that can occur in latch designs. It should be noted that the concept of latching is extremely intertwined with the circuit style(s) chosen (see Chapters 2 and 3), the clocking system (see Chapter 6) and the overall cycle-stealing (if any) approach (see Chapter 8).

The performance of latch circuits can be broken down into several components. The first is just the propagation delay from the data input to the data output with the clock set to a mode that makes the latch transparent. In this mode the latch behaves as if it were a logic circuit and its performance can be optimized as such. The next component is the time the data must arrive before the clock closes the latch such that the data can be safely stored for a sufficiently wide variety of process shifts. The third component is delay introduced into the path by the arrival times of the clock edges. The clock arrival time is the most critical component of a latching system for many reasons, most of which are discussed more completely in Chapters 6 and 8. A well designed latching system allows the most critical latches to become transparent at the optimal time reducing the delay of the latch to just the input to output propagation delay, which is as efficient as any latch can ever be.

The power consumption of a latch is derived from two main components, neither of which have much to do with the current supplied to the actual the storage elements. The first component is the power required to switch the chip-wide clocking network. For microprocessors, clock power can account for 25% of total chip power. [5.1]The second component is the power consumed by switching activity initiated by invalid signal edges passing through a latch. Some latch designs allow any edge that arrives at its input to pass through to the output for as much as one-half of the clock cycle when the valid logic state is finally stored. These edges waste power in the downstream logic by causing invalid switching activity. Any serious attempt to reduce the power consumption of the latches always addresses one of these two issues.

While testability is not an issue addressed by this text, it is important to note that latch circuits must operate in broader range of process and environment conditions. Most logic circuits are not required to function within the system architecture at very high V_{DD} and temperature. With latches, however, this is seldom the case. Even when not required to serve any system function, latch circuits are usually being exercised by a test function. This can range from simply maintaining the current state of its storage node to controlling the behavior of the logic being tested. A latch circuit's test functions can be so important or complicated that the design of the remainder of the latch can be greatly affected. The reader is encouraged to seek other sources of information on testing before implementing a sophisticated latching strategy.

High speed CMOS designs require that latches and their clocking be integrated into the logic circuit design to a much higher degree resulting in latches serving many more functions than simply storing a logic state.

1. Cycle Boundaries: In almost every design, latches are used at the cycle boundaries to pass valid logic states from one pipeline stage to the next.

2. Mid-Cycle Boundaries: The insertion, or movement, of latches to some intermediate point between cycle boundaries is a key element of cycle stealing, a concept which is thoroughly discussed in chapter 8.

3. Execute logic function: One method of reducing latch delay is to merge it with logic function from either its input or output side.

4. Transition from one circuit style to another.

   - Dynamic to Static: Whether as a separate circuit or integrated into a logic circuit, some latch designs are capable of preventing the precharge state of a dynamic signal from propagating to downstream logic without impeding the evaluate state.

   - Single-rail to Differential: An efficient means of creating a differential signal pair is to use the clock as a strobe to simultaneously launch both sides of a latch to the differential outputs.

   - Pulse to Static and Static to Pulse: Signals whose values don't remain valid for a duration set by the clock are said to be *pulsed* signals. Latches are very convenient circuits for transitioning either to or from pulsed logic styles.

5. Initiate test of a portion of the logic: Latch and chip testability will not be discussed in this text.

5.2 Basic Latch Design

Transmission Gate Latch
Tristate Inverter Latch

5.2.1 Storage Elements

A *storage element* is a circuit that holds the state of a signal after the logic that set the signal has changed its state. A *latch* is defined as a single-bit storage element that can store only a single value in response to a control signal. Historically, this type of circuit is typically referred to as a D-latch. From the viewpoint of the microarchitecture, the primary function served by a storage element is to hold the current valid state at the beginning of the logic pipeline while preventing the next valid state from entering the logic pipeline too soon and corrupting the current result. Synchronous designs make use of a system clock or a derivative of the system clock to synchronize the transfer of valid states through the storage elements and across the pipeline boundaries. So in their most primitive form, storage elements are composed of 1) a clock gate that controls when the next valid state is stored and 2) a storage node, a capacitor usually with cross-coupled transistors to maintain the charge, that holds the current valid state at the output while it is processed by the next stage of the logic pipeline.

A *level-sensitive storage element* is one that passes data from the input to the output when the control signal is active. Therefore, the data stored in the element is the last value present in the storage element when the control signal is inactivated. The single level-sensitive latch is the basic building block for all latches and flip-flops. Its logic diagram in Figure 5.1 is that of a multiplexor where the clock selects either D, the next logic state, or Q, the current stored logic state.

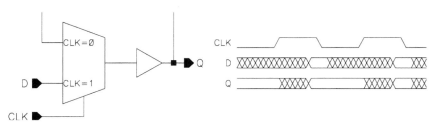

FIGURE 5.1 D-Type Latch with Waveforms.

A *flip-flop*[1] is defined as a single-bit storage element that can store two values in a master-slave type of arrangement in response to a control signal(s). Figure 5.2 shows a simple logic diagram of two D-latches in series. The first latch, also know as the *master latch* and sometimes referred to as the *L1*, stores the value of the input (D) on the rising edge of the clock (CLK). While CLK is low, data propagates up to the internal node (QL) between the two latches. The rising edge of CLK deselects D at the input Mux and selects QL, thus *capturing* the value of D into the master latch. The second latch, also known as the *slave latch* and sometimes referred to as the *L2*, passes, or *launches*, the current value of the master (QL) to the output (Q) also on the rising edge of the clock. The output is held at this level until the next rising clock edge. Because data is both captured at the input and launched to the output by the rising edge of the clock, this circuit is referred to as a *positive edge-triggered flip-flop*.

An *edge-triggered latch* is one that *samples* the input data on either a rising or falling edge, as defined by the circuit, of a control signal. Therefore, the data stored in the element and present on the output is the value that was present on the data input when the control signal made the described transition. In this text, sampling is differentiated from capturing in that sampled data does not enter the latch until the activating edge

---

1. In industry today, much confusion has occurred because many engineers use the term "latch" to mean both a latch and a flip-flop and, in many cases, a register as well. To alleviate this problem, the terms as set forth in this section will be adhered to as best possible. However, rest assured that this delineation will not be followed by many engineers in industry.

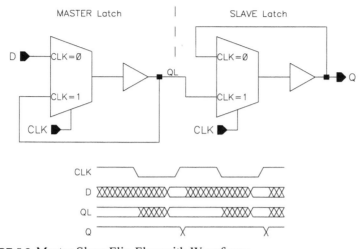

FIGURE 5.2 Master Slave Flip-Flop with Waveforms.

of the control signal. Data that propagates into the storage element before the activating clock edge closes the latch is said to be captured.

5.2.2 Static and Dynamic Latches

The storage node in a latch is simply the capacitor that holds the value sampled from the data input. There are two types of storage nodes used in latch design. A *static latch* uses a storage node that always has a conducting path to either a power or ground connection. A *dynamic latch* has a conducting path to power or ground only while the data is being sampled. Figure 5.3 shows examples of each type.

The static latches have their storage nodes (QA and QB) held by an inverter which completes a non-inverting circuit loop to maintain the charge even when the CLK is low. If the feedback inverter is large enough to overwhelm any leakage currents, then the design allows the latch to hold either stored state indefinitely. A drawback is that when the clock switches high (CLK=V_{DD}), there is current provided by the feedback inverter that is always working against the devices that are trying to switch the storage node resulting in a small loss in performance. Removing this feedback inverter creates a structure known as a dynamic latch, which relies on the capacitance of the storage nodes (QC and QD) to hold the charge until the arrival of the next clock edge. This makes the latch faster but also creates new problems. If the clock does not allow the storage node to be refreshed for a long period of time, noise sources can lower the charge on the gate of the inverter past its switch point.

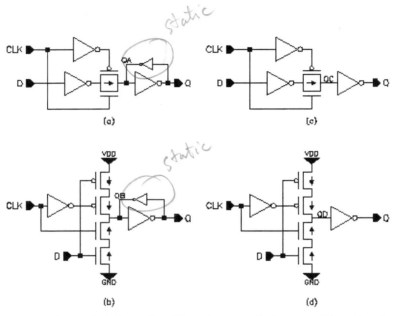

FIGURE 5.3 D-Latch Schematics: (a) static transmission gate, (b) static tristate inverter, (c) dynamic transmission gate (d) dynamic tristate inverter.

5.2.3 Latch Clocking

A primary aspect in the formulation of a latching strategy is the trade off in complexity between the design of the latch circuits and the clocking used to control the latches.

Clock Phases and Edges

The state of a latch is determined by the phase of its clock input. Transparent and edge-triggered latches are considered active when their clock transfers the data input to the data output. An *active high* latch is active when its clock is high or V_{DD} and *active low* latch is active when its clock is low or ground. The *capture edge* of a clock arrives at the end of the transparent phase and causes the value at the data input to be stored in the latch. The subsequent clock level makes the latch *opaque* at the data input. The *launch edge* of a clock arrives either at the beginning of the transparent phase or when data is sampled and causes the valid state to propagate to the data output. For level sensitive latches, the subsequent clock level makes the latch *transparent* at the data input.

Timing: Setup and Hold

As a logic element, latches keep pipeline stages separate by preventing signal edges from propagating into the next stage of the pipeline until the next clock cycle has begun. As a circuit element, a latch is capable of this only if its data and control inputs behave within certain timing specifications.

A signal level must be valid, or *setup*, at a specified time with respect to the capture edge of the clock and that level must be *held* for another specified amount of time with respect to the same capture edge of the clock. The first of these timing specifications is referred to as the *setup time*. Inputs that satisfy the setup time requirement will be latched properly by passing through the latch's clock gate before the capture edge of the clock cuts off the storage node from the input. Failing a minimum setup time requirement prevents the next valid pipeline result from being latched and passes a bad logic state into the next stage of the pipeline.

The second timing specification is referred to as the *hold time*. Passing the hold requirement means that any transition of the data input while the latch is opaque will not affect the state of the latch. Inputs satisfy the hold time requirement by not switching within the timing window that starts with the last edge that meets the setup requirement and the first edge that meets the hold requirement. Failing the minimum hold requirement allows a signal transition that belongs to the next clock cycle to pass into the next pipeline stage one cycle early. This failing condition is referred to as a *race condition*. Consider two consecutive latches where the first latch launches data with a clock edge that arrives at a time similar to the clock edge that captures data in the second latch. The second latch must have captured its data before the output of the first latch arrives and invalidly races through the second latch one cycle too early. This is checked by making sure the earliest arriving data into the second latch does not violate the hold time specification for that latch design.

Effects of Clock Skew

The arrival times of clocks to consecutive latches dictates the behavior of the logic in between and around the latches. *Clock overlap* is defined as to two different clocks causing all of the latches they control to be transparent at the same time. The positive effect of overlap is that data is launched sooner and captured later so more time is available for the data to propagate through the logic. Overlap also causes the logic in between latches to have a minimum delay to guarantee that hold times are met. *Clock underlap* is defined as to two different clocks causing the latches they control to all be opaque at the same time. Its positive effect is to limit, or even eliminate, race conditions. Underlap also reduces the amount of time available for the logic function to complete.

Many of the reasons for requiring overlap or underlap in a multiple clock design have to do with clock skew. The sources of clock skew are covered in other chapters in this text. The effects of clock skew appear in the increase or decrease of the overlap or underlap that was designed into the chip clocking. These effects must be accounted for while verifying that a design has no race conditions affecting chip functionality. For the previous example of two consecutive latches, the first latch must use the earliest launch edge possible and the second latch must use the latest capture edge possible. The resulting hold time check would use arrival time compensated by clock skew for both an early data edge and a capture clock edge. These early arrival times are not always easy to determine as most timing methodologies are designed to work with late arrival times used for calculating chip cycle time. Latch designs that will be detailed later in this chapter have been developed to alleviate this problem by eliminating these race conditions logically.

5.2.4 Noise/Robust Design

Latch circuits are subject to the same noise concerns as any other circuits with weakly held internal nodes. It is obviously imperative that the storage node in a latch be able to hold its charge such that the stored logic value is preserved. When speed is a priority, however, compromises are made that negatively affect the noise immunity of the latches. These compromises must be validated as safe using extensive circuit simulations covering the entire process window. This is almost always worth the effort as almost every improvement in latch performance improves the cycle time of almost every pipeline stage in a design.

DC Noise Sources

Leakage Current: The design of the feedback inverter plays an important role in the noise immunity of the latches in Figure 5.3 (a) and (b). Without a feedback inverter, the speed at which a latch switches is very high but leakage currents can eventually deplete enough charge from the storage node and cause the latch output to switch. In the dynamic transmission gate latch in Figure 5.3c node QC is separated from V_{DD} or ground whenever CLK is low, which is approximately half the time. If QC is high and latch input D has switched low while CLK is low, sub-threshold current through the transmission gate will start to pull QC below V_{DD}. Given enough time, QC will fall below the switch point of the output inverter and the latch output will make a logically invalid switch to V_{DD}. It is clear from this example that there is a minimum refresh rate for which a dynamic latch can be guaranteed to function properly.

Voltage Divider: With a feedback inverter, leakage currents are no longer a concern. Saturation currents for very small devices are strong enough to offset sub-threshold currents for very large devices at even the most severe operating environments. The difficulty is in not making the feedback inverter so strong that the circuit driving the storage node is incapable of sinking both the feedback current and the stored charge. In the static transmission gate latch in Figure 5.3, if QA is high and D is high when CLK rises, the conducting path from latch QA to ground must have low enough resistance to keep the resulting voltage drop well below the switch point of the inverter at the latch output. If the feedback current or the resistance in the path to ground is too high, the latch will certainly be too slow for short cycle times and could even fail to switch at any cycle time. This is why some designs use a tristate inverter to produce a *clocked feedback* design where the feedback inverter is shut off whenever the clock is controlling the latch to open to new data. The only negative aspect of this design is the additional clock load. More clock load leads to more clock skew, a concept that is described in Chapter 6.

AC Noise Sources

Input Noise: For the two transmission gate latch designs in Figure 5.3, the data input can potentially switch the latch while CLK is low. If the source input either drops below ground or rises above V_{DD} by more than the threshold voltage of the clock gate transistors, either the n-device or p-device will turn on and the stored value in the latch could easily be lost. This is most likely to occur when the latch and the circuit driving its input are placed too far apart for the driving circuit to control the coupling noise at the latch input. This type of noise error can only be prevented by the proper design of the devices and wires on the input side of the clock gate.

Charge Sharing: For the tristate inverter, if the data is connected to the gate nodes of the n-device or p-device whose drain nodes are both connected to the storage node, charge redistribution can cause the latch to inadvertently switch. This is a much greater problem for dynamic latches as there is no mechanism to resupply charge to the storage node after CLK is low. This is never a problem for the designs as shown in Figure 5.3.

5.2.5 Latch Implementation

The actual transistor structure of the clock gate and storage node is very closely tied to the circuit implementation of the logic at the input and output of the latch. One example of a transmission gate latch that has a customized implementation is in Figure 5.4. The two latches are connected in parallel and their input and output inverters are replaced with static CMOS logic gates.[5.2] This customization removes two inverter delays from each path reducing the latch delay to only a transmission gate. This type of latch must be implemented in such a manner that the circuit designer has a high level of control over the layout of nodes N1 and N2. The cross-coupled inverters that hold these storage nodes for one half a clock cycle are very weak and if the latch and logic gates are designed as separate elements, there could be excessive bad capacitance on N1 and N2 that increase the likelihood that a noise event will flip the state of either latch.

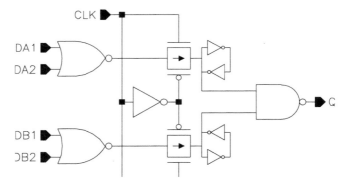

FIGURE 5.4 Logic implemented within two parallel transmission gate latches.

5.3 Latching Single-Ended Logic

Pseudo Inverter Latch (PIL)

True Single Phase Clocking(TSPC)

Double-Edge-Triggered Flip-Flops (DETFF)

The most common type of latch, even in high speed CMOS design, is one that stores single-ended logic inputs. This type of latch generally has the most serviceable timing requirements, does not load the clock excessively and in most forms can receive inputs from logic of most any circuit design style making it an essential component in any chip's circuit library. The t-gate and tristate inverter latches whose function and behavior were detailed in the previous section are the most prevalent versions of latches for single-ended logic. The latches that follow are only a sampling of the very rich set of options for latching single-ended data. They serve to complete the picture of basic latch design started in the previous section.

5.3.1 Pseudo Inverter Latch

The pseudo inverter latch has been described as a component of a high speed latching strategy several times in other published work.[5.3][5.7] It is similar to the tristate inverter latch except it uses fewer transistors and can account for less load on the clock at the expense of tristated nodes under certain input conditions.

Function

Pseudo inverter latches are always composed of at least two stages. Referring to the latch in Figure 5.5(a), whenever the clock (CLK) is high, devices 1, 3, 4 and 6 form a buffer from data (D) to the latch output (Q) making the latch transparent. While CLK is low, devices 2 and 5 are off and block the paths of nodes N1 and Q to ground making the latch opaque. A rising edge on D is blocked by a low CLK in the first stage and causes node N1 to be tristated high. If Q is initially high while CLK is high, a falling edge on CLK will cause Q to be tristated low. A falling edge on D pulls N1 high which is blocked by a low CLK in the second stage and causes Q to be tristated high. The rising edge of CLK will then launch either of these values of D to Q. In this specific example, the tristate condition for N1 is corrected with the addition of device 7, a weak p-device that holds N1 high when CLK is low and D is high.

One of the advantages of this type of latch is the ease with which static CMOS logic gates can be inserted into the first stage. In Figure 5.5(b), a 2-input AND function is merged with the latch. NAND function can be achieved with the addition of an inverter at the latch output. OR and NOR functions can be implemented in the same fashion. Their design can be either dynamic, as the diagrammed example is, or the output can be tied to the gate of a p-device to hold the internal node high. There exists a special case for the design of a this latch when the logic function is a NOR. Devices 1 and 2 in the NAND example would be in series between V_{DD} and the N1. Whenever CLK is low, and N1 is tristated high, inputs D1 and D2 could switch such that charge sharing could drop the voltage on N1 far enough that the output could switch. This can be remedied by controlling the gate of the weak p-device with CLK to hold N1 high. This p-device can replace the charge that coupled into the internal node of

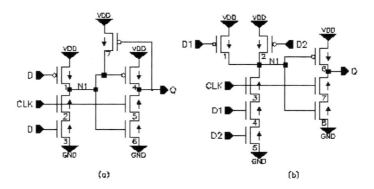

FIGURE 5.5 Pseudo Inverter Latch: (a) D-Latch, (b) D-Latch with 2-input AND function.

the stack without being so large that a falling CLK destroys the high output before device 7 turns off.

Characteristics

This style of latches works quite well for cases where the capacitive load on Q is small. For loads that are larger and especially those that are wire dominant, noise is a concern. There is the possibility that either of the cases where the level of output Q is tristated are susceptible to coupling noise. An extra output would eliminate this problem at the potential expense of extra delay.

5.3.2 True Single Phase Clocking (TSPC)

The latches in Figure 5.5 are transparent whenever CLK is high. Their compliment would be latches that are transparent whenever CLK is low and would differ only in that the CLK is connected to the gate terminal of p-devices. Latches connected in series such that they alternate between active high and low can be controlled by the same global clock without introducing any race conditions or limiting the cycle time with extra clock skew. This is the principle behind *True Single Phase Clocking*. If the same clock edge controls every latch, there is almost no concern of clock skew creating extra overlap or underlap in the clock phases of consecutive latches. With no overlap of clock edges, data can't race through a latch to be captured prematurely by the next latch. Without additional clock underlap, there are no early capture edges reducing the time available for data to propagate through the logic.

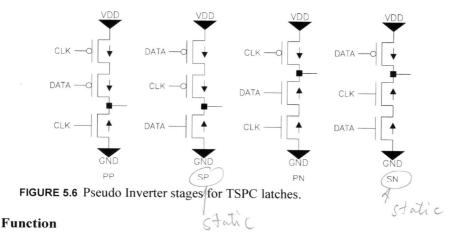

FIGURE 5.6 Pseudo Inverter stages for TSPC latches.

Function

Figure 5.6 shows a set of inverters that can be mixed and matched to form a variety of latches that can be used in a TSPC design. Two p-stages are required for an active low latch and two n-stages are required for an active high latch. A positive edge triggered flip-flop is composed of SP + SP + SN + SN stages for static data input. For pre-charged data input, the same latch is composed of PP + SP + PN + SN stages. In both cases, the master latch is composed of two p-stages that are transparent while CLK is low. The slave latch is composed of 2 n-stages that are opaque while the master latch is transparent and launch the master data to the output when CLK transitions high. Table 5-1 shows the variety latches that can be created by combinations of the p- and n-stages.

p- and n- stages	Input Data	Latch/Flip-Flop Type
SN + SN	Static;	Active High CLK; Level Sensitive
PN + SN	Precharged: Active High	Active High CLK; Level Sensitive
SP + SP	Static;	Active Low CLK; Level Sensitive
PP+ SP	Precharged: Active Low	Active Low CLK; Level Sensitive
SP + SP + SN + SN	Static	Positive Edge-Triggered
PP + SP + PN + SN	Precharged; Active Low	Positive Edge-Triggered
SN + SN + SP + SP	Static	Negative Edge-Triggered
PN + SN + PP + SP	Precharged; Active High	Negative Edge-Triggered

TABLE 5-1 TSPC latches.and flip-flops

Characteristics

Consider the case of a data path that is constructed of an active low latch followed by logic terminating at an active high latch followed by logic terminating at an active low latch. There are several positive aspects of controlling the latches in this configuration with a single clock.

- Because the active high latches are never transparent at the same time as the active low latches, the delay of the logic between latches can be nearly zero. This is because there is almost no clock skew for latches that are in close proximity of one another. In a multiple clock system, the design of the clocks and/or clock skew create overlap in the transparency of opposite phase latches. A significant amount of design time has to be spent adding delay between latches that have no intervening logic.

- The maximum propagation delay through either logic block is one clock cycle. (explained more fully in Chapter 8. This period of time starts with the launch edge of one latch and ends with the capture edge of the next latch. Since both latches are being controlled by the same clock net, the arrival times of the clock to each latch are virtually identical. For the ideal case of a single global clock without jitter, the capture edge arrives at a latch exactly one cycle after the launch edge arrived at the previous latch with no time lost due to clock skew.

There are also a few negative aspects of routing a single clock to control these particular latches in a datapath.

- There are fewer options for routing a single clock across a chip. Each electrically distinct inverter in each buffering stage of the clock driver increases the skew to the point that the single clock now behaves more like it were a multiple clock system. The difficulty with this is that the single clock does have as many options when it comes to negating the clock skew. With a multiple clock system, there is the option to either overlap or underlap the transparency of the clocks to either speed up the slower logic or remove the race conditions from the faster logic. Probably the most efficient method for routing a single clock is to use a single large inverter to drive every functional clock pin on the chip.[5.3] The only source of clock skew would be from the RC between latches in the large clock wiring grid, which would be minimal for latches near one another.

- The slew rate of the single clock has a greater effect on latch design than the slew rates of multiple clocks. The only way data can race through a properly designed master-slave flip-flop that uses a single clock is if the clock slew rates at the latch are large. Using multiple clocks, there can be separation between the clock edge that opens the master and the edge that closes the slave. Since neither of these edges are critical to the cycle time, their slew rates can be a little longer without any negative impact on the operation of the flip-flops. For a single clock, both the edge that opens and the edge that closes the slave latch have critical slew rates. Both edges must be as fast as possible to guarantee both fast and safe operation of the flip-flop.[5.8]

5.3.3 Double-Edge-Triggered Flip-Flops (DETFF)

The concept of having a latch output evaluate on both edges of the clock, called a Double-Edge-Triggered flip-flop, has been evaluated several times in published work. In a conventional flip-flop, capture and launch occur on the same clock edge, making the opposite clock edge superfluous to the operation of the latches. The advantage of a DETFF is that it requires half the clock rate while operating at the same data rate because it can capture and launch data on either edge of the clock. This results in a large savings in the power dissipated by the clock distribution. The challenge is to design a latch with high performance that does not add too much gate load to the clock net.

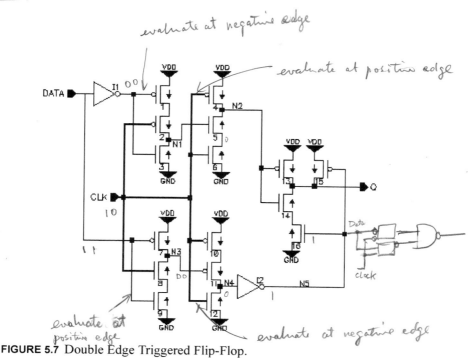

FIGURE 5.7 Double Edge Triggered Flip-Flop.

Function

One example of a DETFF is shown in Figure 5.7. This variety was selected over several other designs for its combination of speed, clock load and robust design.[5.4] Other designs that were faster or loaded the clock less had internal nets that were not always at V_{DD} or ground which might affect the safe operation of the dynamic latches.[5.5]

This DETFF is composed of two parallel latches. Devices 1-6 form an active high dynamic latch and devices 7-12 form an active low dynamic latch. While one latch is launching data that was captured one full cycle previous, the other is capturing data that will launched one full cycle later. When the one latch is passing valid data, the output of the other is precharged to a consistent state. Node N2 is high whenever the active low latch is launching its stored data. Node N4 has complimentary behavior and can be inverted to match N2. The insertion of inverter I2 in the path for the active low latch requires inverter I1 be inserted in the path for the active high latch. Output data is merged with a static NAND gate.

Characteristics

The advantages of double edge-triggering exist primarily in the power savings realized when halving the clock frequency. It may also be possible to use this technique in local areas of the chip as well. Some chip operations are required to take place twice per cycle. Instead of rerouting a separate doubled clock, the logic executing such function could be bounded by DETFFs.

The decision to use double edge-triggering for any latch design has a significant impact on the design of the logic circuits. Clocked circuits cannot be used along with DETFFs without doubling the clock frequency for the logic, which would defeat the advantage of this latch style. It is very likely, however, that if low power design is such a high priority that DETFFs are used, static logic will probably be the most attractive logic circuit style.

DETFFs do not allow any cycle stealing techniques to be employed. This includes any overlapping of clock phases that can be used in other master-slave latches that receive multiple clocks. The clock must have as near a 50/50 duty cycle as is possible. Any extra uncertainty in one clock edge subtracts from the minimum attainable cycle time for a design. These flip-flops can only be used in pipelines in which non-clocked logic is bounded by master-slave latching.

5.4 Latching Differential Logic

Differential Cascode Voltage Switch (DCVS) Latch
Static RAM Latch
Ratio Insensitive Differential Latch
Differential Flip-Flops

Differential logic circuits can be modified such that they function as a latch with very little difficulty. Typically, there is already some form of cross-coupled relationship between the true and compliment logic function in the circuit. Very little and sometimes even no modification is required to make the circuit behave as a latch. How much modification is required usually depends on the operational requirements of the chip such as its minimum clock rate, high voltage and high temperature tests.

5.4.1 DCVS Latches

The DCVS latch[5.6] described in this section is very similar to the logic circuit from Section 3.3.1. The difference is the clocked device that causes the latch to be opaque when CLK is low making the latch active high. This circuit will be the basis for the design of the remainder of the latches in this section.

Function

In Figure 5.8, devices 1 and 2 represent, respectively, the true and compliment logic function. One of the two sides will pull the output to ground while turning on the cross-coupled p-device that pushes the other output high.

Characteristics

In addition to the features of DCVS detailed in Section 3.3.1, an aspect of this latch that needs to be closely analyzed is the dynamic storage node. While CLK is low, the n-tree that evaluated has its output tristated low. Obvious potential hazards in this case are leakage raising the storage node to V_{DD} or the output node coupling either to edges on the data inputs or somewhere along the output net. Protection from most coupling events can be achieved at the cost of delay by adding an inverter to the output thereby limiting the physical size of the storage node and its interaction with external signals. If neither of these is a problem, then the DCVS circuit shown is a very fast and dense latching solution.

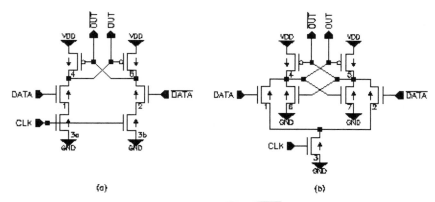

FIGURE 5.8 Differential input latches: (a) DCVS, (b) Static Ram.

A designer might be tempted to merge devices 3a and 3b since only one of either device 1 or 2 will be conducting.   For a latch, this presents a potentially serious problem. With only one clocked device, the source terminals of devices 1 and 2 are connected. While CLK is low, the latch must be opaque to any change at the data inputs. At the time the data is switching, both devices 1 and 2 are on and there is a "sneak path" where current can flow between OUT and $\overline{OUT}$. Whichever node was being held high can probably recover, but the output that must remain low has no path to ground and any charge that accumulated while the data inputs were in transition remains leaving the node at a potential higher than ground. Even if the DATA signal pair transitions very quickly, charge (V_{DD}-V_{TN}) that has accumulated at the drain node of the clocked n-device will be shared with the output node that is being held at ground creating a poor logic level at the output.

5.4.2 Static Ram Latches

The static RAM latch[5.6] is a simply a 6 transistor SRAM cell with the addition of a clocked device in the path to ground that controls when the latch is transparent.

Function

The addition of the cross-coupled n-device does nothing to change the operation of the static RAM latch from that of the DCVS latch.

Characteristics

The static RAM latch overcomes the problem with the DCVS latch that while CLK is low, one of the outputs is tristated low. The storage nodes are always held at their high or low level. There is a performance penalty for adding the cross-coupled n-devices. For an output to switch high, one n-tree must overcome current from a weak p-device, just as DCVS circuits must do. For the static RAM latch, when the n-tree is low and the opposite side cross-coupled inverter switches, there is switching current that is not present in the DCVS design. So both transitions through a static RAM latch have switching current while in DVCS only one of the two transitions does.

5.4.3 Ratio Insensitive Differential Latch

Both the DCVS and static RAM latch have switching current that affects their operation. Similar to the feedback current in the transmission gate and tristate inverter latches, this current from the cross-coupled devices must be overcome by the input circuit in order for the latch to switch. This at least increases the time required to set the latch and at worst can cause the latch to fail to switch at one or more process corners.

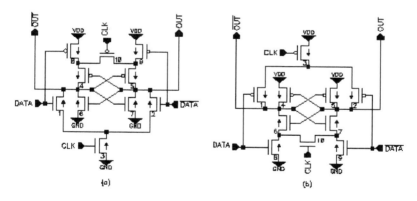

FIGURE 5.9 Ratio Insensitive Differential Latch: (a) active high, (b) active low.

Function

The static realization of a ratio insensitive differential latch is shown Figure 5.9.[5.7] The first schematic shows the latch in its active high implementation. It is the same as the static RAM latch with the addition of devices 8, 9 and 10. Devices 8 and 9 disconnect the path to V_{DD} for the cross-coupled inverters. When DATA is high, devices 1 and 9 will be on and devices 2 and 8 will be off. When $\overline{DATA}$ switches high, device 9 will cut off the current from V_{DD} and allow OUT to be pulled low with minimal switching current. DATA simultaneously switches low and with OUT also low, devices 4 and 8 pull $\overline{OUT}$ high. Device 10 is used to keep the high side of the latch from tristating while the CLK is low and the latch is holding its outputs. While the latch is opaque, the data inputs can switch which turns off either device 8 or 9, whichever had been holding its output node high. Device 10 bypasses these devices while only supplying charge to the output node that is being held high.

Characteristics

The p-type version of this latch in Figure 5.9b is far superior to the p-type DCVS and static RAM latches. For an n-latch, it is a fairly simple to design the proper ratio of n-device logic versus weak p-device feedback. For the p-latch, however, making the n-device weak enough so that the p-device logic can overwhelm the feedback current across the entire process window while still being fast is very difficult. This can make using the DCVS and static RAM p-latches with any imbedded logic impractical. This is not the case for the ratio insensitive latch. Not only is the p-latch without imbedded logic faster than either of the other two types[5.7], but, since there is no device strength ratio to design for, adding logic can be done without a drastic increase in device sizes.

There is increased complexity, however, when imbedding logic into a ratio insensitive latch. For the n-type latch, any logic function that replaces n-devices 1 and 2 must also be implemented with p-devices that replace devices 8 and 9. For AND function, device 1 would be replaced with n-devices in series and device 2 would be replaced by n-devices in parallel. for the pull up logic, device 9 would have to changed the same as device 1 and device 8 would have to be changed the same as device 2. This holds true for p-type latch as well.

5.4.4 Differential Flip-Flops

The characteristic of DCVS circuits that makes then so fast with such high logic density can also be utilized to make fast and dense flip-flops. Typically, a master latch won't pass any input transitions to its output while opaque. For a DCVS slave n-latch, falling transitions at its input won't switch the output regardless of which state the latch is in. First, this simplifies the design of the master latch because falling outputs don't need to be gated by the clock. Second, and most importantly, the master can be an n-latch even when controlled by the same clock as the slave.

Function

The differential flip-flop diagramed in Figure 5.10 is positive edge triggered with a dynamic master latch and a static slave latch.[5.7] The slave latch is a static RAM n-latch identical to the one in Figure 5.8. The master latch is a derivative of the DCVS n-latch with the only change involving how the clock gates the data. Since the master latch needs to be transparent while CLK is low and the slave latch is unaffected by falling inputs at any time, only the pull-up path needs to be cut off by CLK while the latch is opaque. A transition of the data input while CLK is high will cause both L1 and $\overline{LT}$ to be ground. As long as the data held its level after the rising CLK for long enough to safely latch the valid data in the slave, the falling inputs to the slave will not affect the flip-flop outputs.

Characteristics

The strengths of differential latching can be seen when this flip-flop is compared to a fully static pseudo inverter flip-flop. The pseudo inverter implementation uses 14 devices, four of which load the clock, to produce a single-rail output. The complement output requires an inverted true output. This can be done, with minor modifications, within the same 14 transistors The differential flip-flop only loads the clock with two devices and is faster even without imbedding extra logic function in the master latch. Since differential circuits have superior logic density, the performance improvement will only be greater as logic is added to the flip-flop.

It may be desirable to make the master latch fully static, just as the slave latch already is. The inset in Figure 5.10 shows the additions necessary to do this. Since a noise event that causes a master latch output to transition low has no effect on the slave, it is only necessary to prevent the storage nodes from switching high while CLK is high. During this time, DATA can switch high pulling $\overline{LT}$ low and tristating L1 low. Device 4 turns off and device 5 turns on sharing the charge stored on node CVDD and raising the voltage on node L1. The addition of the minimum sized inverter and two n-devices provides this charge with a path to ground. As device 1 pulls $\overline{LT}$ low, it also begins to pull CVDD low and the inverter turns on the new n-devices which fully dissipate all of the charge on CVDD. When CLK transitions low, the inverter turns off the n-devices and functions normally.

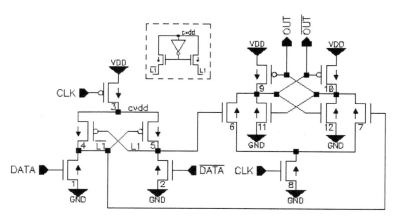

FIGURE 5.10 Positive Edge Triggered Differential Flip-Flop

One negative aspect of the implementation of the master latch is that the setup time can become slightly longer when compared to a DCVS p-latch. A rising data input has to discharge not only the master latch output but also the shared node CVDD. This is less the case as logic function is imbedded as the n-tree of the master latch is superior to the p-tree of a DCVS p-latch in both logic speed and density.[5.7]

5.5 Race Free Latches for Precharged Logic

Cross-Coupled Differential Output

Negative Setup Pipeline Latch

Controlling precharged circuits with a single clock presents two challenges to their safe and speedy operation. First, the latch in between precharged circuits of opposite phase must become opaque before its inputs are precharged. Precharging and capturing the data are initiated by the same clock edge resulting in a race between the closing of the latch to capture the result of the evaluate phase and the precharging of that result. Second, using a single clock to delineate the evaluate phases is affected by a non-50/50 duty cycle. This uncertainty impacts performance by shortening the length of each evaluate phase.

5.5.1 Cross-Coupled Differential Output

One technique for controlling dynamic circuits with a single clock is to precharge circuits while the clock is high and predischarge circuits while the clock is low. Additionally, the use of differential logic allows cross-coupled gates to be used as latches for the outputs that drive circuits of opposite phase. The differential data has the quality of switching the latch with its evaluate data while also blocking the precharge data regardless of its arrival time.[5.10]

Function

Figure 5.11 shows two precharged circuits of opposite phase each having their outputs latched by cross-coupled gates. The first circuit is the evaluate phase when the clock is high and uses cross-coupled NAND gates for its latch. Initially, both precharged nodes are high and the cross-coupled NAND latch is open to receive new data. Before the clock falls, one of the differential outputs will be pulled to ground causing one of the NAND gate outputs to toggle high. The other NAND output will be forced low closing the latch to incoming precharged data. When the clock falls, the differential signals are precharged high. The NAND latch outputs remain the same as the data switch only reopened the latch to receive new data without destroying the outputs. The second circuit functions in the same way, only with opposite logic levels. The circuit evaluates when the clock is low and the latch is open when its differential data inputs are both ground. Its latch is composed of a cross-coupled NOR gates which also close to new data when its inputs are of opposite polarity and reopens when the inputs predischarge while holding the evaluated outputs.

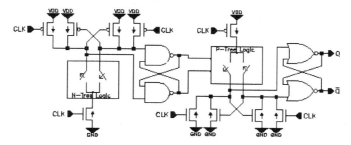

FIGURE 5.11 Pipeline latch for complimentary domino logic.

Characteristics

The latches used in this implementation are actually being controlled by the data inputs and not the clock. They are edge-triggered latches which are closed by the evaluate data and opened by the precharge data. Since the precharge data only opens the latch for the next evaluation without affecting the latch outputs, there is no critical race between closing the latch and precharging its inputs. Additionally, since the clock is not their control signal, these latches do not require the clock to have either of its edge rates be fast. While slow edge rates normally create race conditions in TSPC latches, this hazard is logically impossible for this latch design. It has be shown that these latches function without failure even when the clock input is a sinusoidal wave.[5.10]

The drawback with these latches is that an underlap of evaluate phases is required. Every latch output pair, because they are never reset, behaves statically and has a value of either 10 or 01. Static inputs to a precharged circuit must reach a valid logic state before the arrival of the clock. In order for the next half cycle to evaluate correctly, the latch output must be valid before the beginning of the next evaluate phase. This sets the time available for either evaluation phase to less than one half the clock cycle by the maximum amount of uncertainty in either clock edge. This increases the minimum cycle time for the entire design by at least twice the duty cycle uncertainty.

5.5.2 Negative Setup Pipeline Latch

The difficulty with using cross-coupled NAND and NOR latches as inputs to precharged circuits is their static outputs. Any latch design that allows data to flow through both the latch itself and the following precharged circuits without paying a clock skew penalty must precharge it outputs. Shown in Figure 5.12 is a negative ETL that remains transparent with precharged outputs until after differential data evaluates at the input regardless of the level of the clock.[5.11][5.12]

Function

When CLK switches high, outputs Q and $\overline{Q}$ are both held low and the internal nodes N1 and N2 are both precharged high enabling both n-trees to receive input. Data arriving before CLK falls evaluates through one of the two n-trees and is latched into node N1 or N2. When CLK falls, nodes N1 and N2 are launched to outputs Q and $\overline{Q}$. If CLK falls before the data arrives, the outputs remain predischarged low and the latch remains transparent. The late arriving data propagates through the latch without waiting. The ANDing of nodes N1 and N2 with the input data has the effect of closing the latch to new data until CLK rises again and precharges both n-trees. Because the precharge level of the data is low, the active high n-tree logic can never pass a pre-charge edge to the output regardless of the arrival times of data or clock.

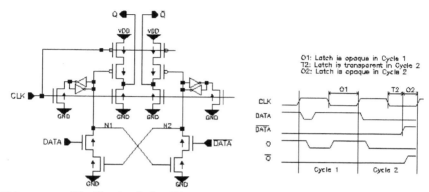

FIGURE 5.12 Pipeline latch for differential precharged logic.

Characteristics

This latch takes advantage of the differential nature of its inputs to eliminate both clock edge uncertainty and setup time from the calculation of the minimum cycle time for a path. The timing diagram in Figure 5.12 shows the two different ways the differential output can evaluate. Cycle 1 diagrams the case when the data arrives before the launch edge of the clock and Cycle 2 has the clock falling before the data is valid.

Clock	Differential Input Data	Latch State	Data Outputs
HIGH	Evaluated to Precharged	Open; Opaque	Precharged
HIGH	Precharged to Evaluated	Closed; Opaque	Precharged
HIGH to LOW	Precharged	Open; Transparent	Precharged
HIGH to LOW	Evaluated	Closed; Opaque	Evaluated
LOW	Precharged to Evaluated	Closed; Opaque	Evaluated
LOW	Evaluated to Precharged	Closed; Opaque	Evaluated
LOW to HIGH	Evaluated	Closed; Opaque	Precharged
LOW to HIGH	Precharged	Open; Opaque	Precharged

TABLE 5-2 Pipeline Latch: Phases of operation

Table 5-2 lists the various states the latch can be in with respect to the state of its inputs. Each row shows the state of the latch and the outputs after the clock or one of the differential inputs switches while the other remains constant. Notice that the only condition is which the latch is transparent is when the data inputs are precharged while the clock falls. When the data does arrive before the clock rises, it flows through the latch and also through the downstream logic which is evaluating while the clock is low. The only other time the latch is open to new input is when the clock is high and the data is precharged.

There are some device ratios that need to be correct for the latch to function while the clock is high. During precharge of the latch, the cross-coupled inverters attempt to hold nodes N1 and N2 high for the entire half-cycle. For data arriving while the clock is low, the n-trees must be strong enough to discharge node N1 or N2 while the cross-coupled inverters are precharging the latch. If they can't, the cross-coupled n-trees will not be able to block the precharge data and a hold time requirement will be necessary for the evaluate data edges with respect to the falling clock. Such a hold time requirement would defeat the purpose of using this latch design.

5.6 Asynchronous Latch Techniques

Self-Resetting CMOS (SRCMOS) Latch

Muller C-element

Asynchronous TSPC Latch

One of the best utilized asynchronous design techniques is cycle stealing. Its basic premise is to not rigidly set the length of evaluation time for any one latch to latch segment in the logic. One segment can use time from a previous segment provided that propagation delay through the two segments adds up to a predetermined maximum time. The advantage is that decoupling the propagation of data from all clock edges can eliminate the need for budgeting clock skew in the delay of many latch to latch paths. Such a design requires latches that are transparent for at least half the clock cycle allowing data to flow through latches and logic without stopping and losing performance waiting for a clock to restart the flow. Almost any of the latches discussed so far in this chapter can be used in logic configured to cycle steal. This section will detail several unique latch designs required by other asynchronous design techniques.

5.6.1 SRCMOS Latch

One variety of self-timed logic circuit is SRCMOS (see Section 3.2.6). These logic circuits transition from evaluate phase to precharge phase and back to evaluate without being controlled by an external clock. Synchronization of SRCMOS circuit operation is most easily realized through the use of an edge-triggered latch that receives either pulse or static inputs and launches a pulse output that initiates the logic wave through a pipeline stage. The remainder of the logic in the pipeline stage can evaluate without external control for one full cycle when it is latched before being relaunched.[5.13][5.14]

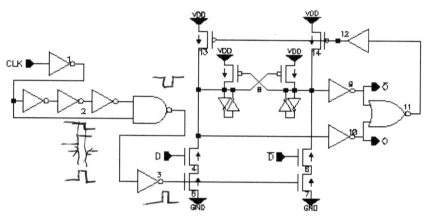

FIGURE 5.13 SRCMOS Latch.

Function

The latch shown in Figure 5.13 starts in a closed state with devices 13 and 14 being off after precharging both sides of the latch high. It uses the falling edge of CLK to sample input data and launch output data. The clock is driven by an odd number of inversions which creates a rising edge into devices 5 and 7. Circuit 2 is a pulse generator composed of three inverters and a NAND gate which transforms the falling edge at CLK into a pulse at the output of inverter 3. The width of the pulse is determined by the delay of the three inverters in the pulse generator. Either data must arrive at D and $\overline{D}$ before the arrival of the clock pulse to the n-trees or D or $\overline{D}$ must evaluate monotonically during the clock pulse. The logic in the two n-trees must be complimentary such that exactly one side of the latch evaluates every cycle. The discharge will turn on one of the two cross-coupled p-devices which hold the opposite side high. Inverters 9 and 10 drive the latched data to the outputs. Two signals logically equivalent to the outputs are input to a NOR gate that switches when either output switches and generates a reset signal for the latch. The reset signal is delayed by enough buffering so that when reset p-devices 13 and 14 precharge the latch and predischarge the outputs, there has been ample pulse width on either Q or $\overline{Q}$ for the downstream logic.

The resetting of the outputs will also switch the state of the reset signal to V_{DD} and turn off devices 13 and 14. If the clock input to the n-trees has not switched low before precharging of the latch begins, the latch output pulse will definitely be wider and could possibly not reset at all. While not shown here, it is possible to interlock the falling edge of the internal clock with the initiation of reset, which would eliminate this problem.

Characteristics

This SRCMOS latch is useful for starting a pipeline of logic circuits that process pulses at some or all of their logic inputs and outputs. As mentioned in Section 3.2.6, the design of pulsed signals such that their active states overlap sufficiently with each other and not with the reset of the circuits that receive the pulses is complicated and labor intensive. The use of an edge triggered latch to generate the pulses at the beginning of a pipeline stage does aid the task of synchronizing the arrival of these pulses at the logic gates. If the first stage in a pipeline receives all of its pulsed inputs from latches, then the variation in arrival times will mostly be due to the clock skew at the latches generating the pulses. This value is small and deterministic enough to allow for a functional design.

5.6.2 Muller C-element

Self-timed circuits often use control signals derived from data signals to indicate that data inputs to the next circuit are ready or to indicate that the next circuit acknowledges receiving or has completed processing the data inputs. These signals are often collected by a circuit called a Muller C-element which behaves as a non-inverting buffer when all of its inputs are the same and otherwise holds the value of the previous time all of the inputs were the same.

Function

For the Muller C-element in Figure 5.14, devices 1, 2, 3 and 4 function as an inverter whenever D0 and D1 are equivalent. The inverter output drives a second inverter composed of devices 5 and 6 which sets the latch and OUT to the same level as the inputs. If D0 and D1 are ground then OUT is ground, and if D0 and D1 are V_{DD} then OUT is V_{DD}. The first inverter output tristates whenever D0 and D1 are opposite and the previous value of OUT is held by the cross-coupled inverters composed of devices 5, 6 and 7.

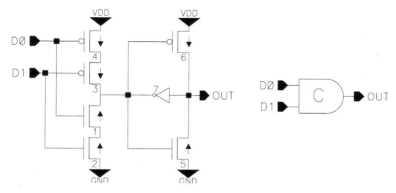

FIGURE 5.14 Muller C-element transistor schematic and logic symbol.

Characteristics

The Muller C-element is primarily used as a controlling circuit for either latches or clocked logic circuits. The c-element inputs and outputs are used as handshaking signals that allow consecutive circuits or pipeline stages to transfer data asynchronously. Their design can range from the simple 2-input circuit in the example to an n-input Muller C-element. Since generation of control signals for logic and latch circuits is usually in the critical path, it is preferable to reduce the number of inputs to the c-element to keep its delay small. The c-element requires processing only those signals whose timing uncertainty might cause them to arrive last.[5.15]

5.6.3 Asynchronous TSPC ETL

Any latch or flip-flop requires an external signal to initiate both the sampling of input data into the latch and the launching of stored data to the downstream logic. Every latch and flip-flop discussed in this text thus far used a system clock to control these functions. Without a system clock to control the passing of data from one pipeline stage to the next, latch design becomes very complicated as shown in this example of an asynchronous ETL.

Function

The negative edge-triggered asynchronous flip-flop in Figure 5.15 uses a Muller C-element output as its clock signal. The C-element processes completion signals from input and output circuits to open and close the TSPC master and slave latches at the proper times. A completion signal is high whenever data has been captured by the master latch and is ground once the master latch has been precharged. The control signals output by the c-elements guarantee that while the slave latch of one flip-flop is set, the master latch in the next flip-flop is also sampling input data. The control signals also concurrently set the previous and next slave-master pair in the precharge phase. Thus consecutive stages of the pipeline are always in the opposite phase of operation just as if they were controlled by the same global clock.

The c-elements in Figure 5.15 show how the completion signals are used to control three consecutive pipeline stages. In_Ready is the completion signal from the latch providing the data input to the first pipeline stage. This signal is low when its master latch has been precharged which also means that the data out of the slave latch and into the pipeline is valid. Prev_Completion is the completion signal from the first pipeline stage and Next_Completion is from the third pipeline stage. The flip-flop shown captures data at the end of the second pipeline and outputs data into the third. While In_Ready is low, Prev_CTRL and Prev_Completion are high and the first pipeline is evaluating. Once In_Ready switches high, pipeline one is in precharge phase and data is launched from the slave latch into the second pipeline. Once precharge causes Prev_Completion to switch high, CTRL to the second flip-flop switches high and enters the evaluate phase ready for the data input propagating through the second pipeline. Once that data arrives, Completion switches high causing Next_CTRL to precharge the third flip-flop which causes Next_Completion to switch low. After the first flip-flop has reentered the evaluate phase, CTRL switches low and launches data out of the slave latch. The precharging of Completion causes the final pipeline stage to evaluate and the data is subsequently sampled at the master latch at the end of the pipeline.[5.16]

5.7 Summary

It is necessary to develop a chip-wide latching strategy for the usage and implementation of latch circuits. Clock routing and skew, race conditions, and timing considerations are problems that must be solved consistently across the chip. Pipeline stages that frequently pass data to each other need a consistent implementation of logic and latch circuits that optimize their performance. Many of the latch designs that have recently appeared in the literature and in commercial products can be solutions to these difficult problems, but only when their use is sufficiently consistent throughout the design. Their use must also be paired with the appropriate circuit styles detailed in Chapters 2 and 3 and with the appropriate clocking methodology, varieties of which will be presented in Chapters 6 and 8.

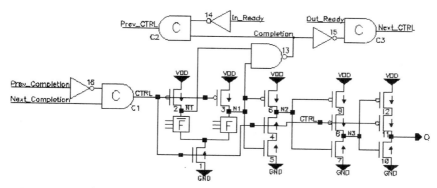

FIGURE 5.15 Asynchronous control of a TSPC ETL.

REFERENCES

[5.1]   P. Gronowski, "Designing High Performance Microprocessors," *Proceedings of the 1997 Symposium on VLSI Circuits*, pp 712-717.

[5.2]   B. J. Benschneider, etal, "A 300-MHz 64-b Quad-Issue CMOS RISC Microprocessor," *IEEE Journal of Solid-State Circuits*, Vol. 30, No. 11, November 1995, pp. 1203-1211.

[5.3]   D. D. Dobberpuhl, etal, "A 200MHz 64-b Dual-Issue CMOS Microprocessor," *IEEE Journal of Solid-State Circuits*, Vol. 27, No. 11, November 1992, pp. 1555-1564.

[5.4]   M. Afghahi, J. Yuan, "Double Edge-Triggered D-Flip-Flops for High Speed CMOS Circuits," *IEEE Journal of Solid-State Circuits*, Vol. 26, No. 8, August 1991, pp. 1168-1170.

[5.5]   J. S. Wang, "A New True-Single-Phase-Clocked Double Edge-Triggered Flip-Flop for Low Power Design," *Proceedings of 1997 IEEE International Symposium on Circuits and Systems,* pp 1896-1899.

[5.6]   N. Weste, K. Eshraghian, *Principles of CMOS VLSI Design*, Second Edition, Reading, MA: Addison-Wesley, 1995, pp 328.

[5.7] J. Yuan, C. Svensson, "New Single-Clock CMOS Latches and Flip-Flops with Improved Speed and Power Savings," *IEEE Journal of Solid-State Circuits*, Vol. 32, No. 1, January 1997, pp. 62-69.

[5.8] P. Larsson, C. Svensson, "Impact of Clock Slope on True Single Phase Clock CMOS Circuits," *IEEE Journal of Solid-State Circuits*, Vol. 29, No. 6, June 1994, pp. 723-726.

[5.9] G. M. Blair, "Comments on 'New Single-Clock CMOS Latches and Flip-Flops with Improved Speed and Power Savings," *IEEE Journal of Solid-State Circuits*, Vol. 32, No. 10, October 1997, pp. 1610-1611.

[5.10] D. Renshaw, C. H. Lau, "Race-Free Clocking of CMOS Pipelines Using a Single Global Clock," *IEEE Journal of Solid-State Circuits*, Vol. 25, No. 3, June 1990, pp. 766-769.

[5.11] C. Heikes, "A 4.5mm^2 Multiplier Array for a 200MFLOP Pipelined Coprocessor", *Proceedings of the 1994 International Solid-State Circuits Conference*, pp 290-291.

[5.12] J. D. Yetter, "Universal Pipeline Latch for Mousetrap Logic Circuits," *U.S. Patent #5,392,423*, June 13, 1994.

[5.13] H. Partovi, etal, "Flow-Through Latch and Edge-Triggered Flip-Flop Hybrid Elements," *Proceedings of the 1996 International Solid-State Circuits Conference*, pp 138-139.

[5.14] W. Hwang, etal, "A Pulse-to-Static Conversion Latch with a Self-Timed Control Circuit," *Proceedings of the 1997 International Conference on Computer Design*, pp 712-717.

[5.15] T. Wuu, etal, "A Design of a Fast and Area Efficient Multi-Input Muller C-element," *IEEE Transactions on Very Large Scale Integration (VLSI) Systems*, Vol. 1, No. 2, June 1993, pp. 215-219.

[5.16] J. Ahmed, S. G. Zaky, "Asynchronous Design in Dynamic CMOS," *Canadian Conference on Electrical and Computer Engineering v. 2 1997*, pp 528-531.

CHAPTER 6

Interface Techniques

6.1 Introduction

The benefits of the high-speed CMOS logic circuitry described in the preceding chapters can only provide system-level performance enhancements if high-speed input and output (I/O) circuit techniques and high-bandwidth interface protocols are also incorporated into a design. Achieving high frequency chip-to-chip communications requires identifying and quantifying the environment in which the I/O circuitry must operate and optimizing the circuitry accordingly.

This chapter examines interface considerations for short-channel CMOS devices in high-speed applications. The intent of this discussion is not to provide in-depth information about power busing, package, or card design, but to illustrate practical approaches for minimizing the impact of these and other environmental effects on I/O circuitry performance. Several useful driver, receiver, voltage translation, and ESD circuit techniques are presented in the context of typical CMOS applications.

6.1.1 System-Level Considerations

I/O system-level considerations may be summarized as:

1. Interface timings, valid signal levels, and signaling protocols
2. Chip-to-chip communication networks
3. Packaging constraints and parasitics

4. Electrostatic discharge (ESD) protection
5. Voltage translation and over-voltage protection
6. Receiver and off-chip driver circuit topologies
7. External and self-induced sources of noise

A good interface design balances trade-offs among these system considerations to achieve the desired performance. At data rates approaching 1GHz and beyond, system noise dominates interface performance if not carefully examined in the initial design of the interface circuitry and package connections. Thus many of the techniques discussed in this chapter are developed with the intent of minimizing the impact of system noise on driver and receiver circuit performance.

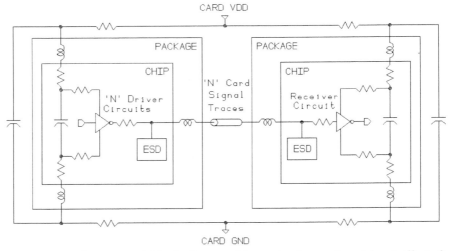

FIGURE 6.1 System model including on-chip, packaging, and card-level effects for I/O optimization.

Figure 6.1 illustrates a general model of an I/O system. Extensive models also include on-chip wire inductance, mutual inductances between package connections, and card-level line-to-line coupling.

Beginning with the driving circuit, an on-chip voltage swing of ground to V_{DD} (internal power supply) must be communicated off-chip to external levels, typically a minimum valid high output level, $V_{OH}(min)$ and a maximum valid low output level, $V_{OL}(max)$. Single-ended output signals are provided individually and are referenced to a supply level common to both the sending and receiving chips (i.e. ground or a reference potential). Differential output signals are provided in pairs which transition about a common-mode voltage.

The simultaneous switching of the output drivers results in self-induced noise as mentioned in Chapter 4. The majority of this noise results from inductance in the power and ground connections through the chip package to the driver circuit. This inductance must be considered when selecting a package appropriate for a particular application. Multiple V_{DD}, V_{DDQ} (driver supply voltage), and ground connections through the package can reduce the effective inductance of these supply paths. The impedance of on-chip power buses, however, can isolate the driver circuitry from some of these paths. Careful chip architecture focuses on placing high-performance I/O circuitry as close to their power supply connections as possible. On-chip power supply decoupling capacitors, common in microprocessors and logic chips, can also reduce driver simultaneous switching noise. This capacitance is only effective, however, if located adjacent to the driver circuitry.

Matching the wire lengths from driver output to chip output pad while avoiding capacitive coupling to other on-chip signals ensures matched timings. The shortest possible paths through the chip package are desirable as well, with lengths again matched for critical output signal timings. The speed of short interconnects must, however, be balanced with the need for ESD protection. Such protection often requires the use of longer, resistive connections in series with driver output signals as discussed in Section 6.4.

The external signals must then traverse some interconnect, whether card traces or multi-chip module (MCM) connections. The signals are subject to additional deformations at this point due to transmission line effects and line-to-line coupling. Section 6.3 discusses transmission line effects. External decoupling capacitors are common, stabilizing the card power supply provided to the chip. For low-impedance card power plane connections, these external capacitors do not reduce the on-chip simultaneous switching noise of the driver circuits.

At the receiving chip, the signals must achieve either a minimum valid input high level, $V_{IH}(min)$, or a maximum valid input low level, $V_{IL}(max)$. The signals are once again subject to packaging parasitics and are then imposed on a receiver circuit. As with driver circuits, receiver circuit performance may be affected by simultaneous switch noise, particularly the switching noise of driver circuits. It is always preferable, though not always realistic, to switch the drivers at some time other than when the receivers are evaluating [6.33]. Guaranteeing such timings at high frequencies is difficult, if not impossible.

6.2 Signaling Standards

CMOS integrated circuits can now be found in every imaginable environment, including PC's, workstations, fiber optic systems, wireless communications, and household appliances. The variety of interface challenges these applications pose is further complicated by the constant reduction of internal chip power supplies as device channel lengths decrease. The Joint Electron Device Engineering Council (JEDEC) has established standards to provide compatible I/O levels among integrated circuits.

JEDEC standard interface level definitions include both small signal swing and full signal swing conventions. Stub series terminated logic (SSTL), high speed transceiver logic (HSTL), and gunning transceiver logic (GTL) all specify small amplitude voltage swings about a reference voltage [6.1][6.2][6.3]. Low voltage CMOS (LVC-MOS), and low voltage transistor-transistor logic (LVTTL) use unterminated, rail-to-rail voltage transitions [6.4][6.5]. Signaling levels in some logic families, including LVTTL and emitter-coupled logic (ECL) [6.6], were originally defined for bipolar or biCMOS output circuitry. CMOS output circuitry can be adapted for compatibility with these interface requirements [6.24][6.25].

The parameters commonly used to specify I/O signal levels are:

V_{DD}: chip supply voltage (also driver supply voltage for LVCMOS and LVTTL)

V_{DDQ}: driver supply voltage

V_{REF}: input reference voltage

V_{TT}: termination voltage

V_{IH}, V_{IL}: valid high and low input levels, respectively

V_{OH}, V_{OL}: valid high and low output levels, respectively

These parameters usually have both AC and DC levels specified. Figure 6.2(a) superimposes these levels (dashed lines) on a valid data transition (solid line).

In these interface definitions, data integrity requires zero slope reversal through the valid signal levels. Slope reversal is undesirable because the correct transfer of data from one chip to the next must satisfy proper set-up and hold times for latching of the data, as described in Chapter 5. The valid data transition of Figure 6.2(a) may be compared to the transition illustrated in Figure 6.2(b). The latter signal can result in the corruption of transmitted data due to multiple excursions through the $V_{IL}(AC)$ level. Such slope reversal is of particular concern in high-frequency communication networks where transmission line effects can produce significant signal reflections and dominate signal integrity.[1] Overshoot and undershoot, also illustrated in

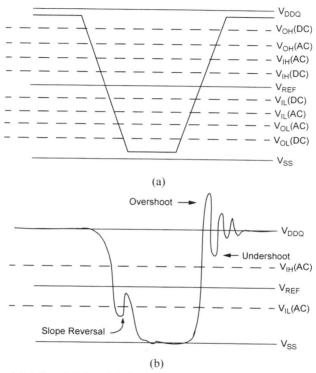

FIGURE 6.2 (a) AC and DC valid signal levels and valid transmitted signal. (b) AC and DC valid signal levels and potential corruption of transmitted signal.

Figure 6.2(b), can cause data corruption and, over time, I/O device performance degradation.[2]

In many applications, both the valid signal levels and the entire interface protocol are standardized. The dynamic random access memory (DRAM) industry provides one example of protocol standardization. Both synchronous link DRAM (SLDRAM) and Direct Rambus DRAM use terminated, small-signal swing interface levels: SSTL and Rambus signaling logic (RSL), respectively.[3] Each standard also defines clocking and

---

1. See Section 6.3.

2. 'Undershoot' is often used to describe a transient below ground and 'overshoot' a transient above V_{DD}. Figure 6.2(b) illustrates these transients as referred to for damped oscillations.

---

control signal strategies with the objective of maximizing memory bandwidth [6.7][6.8].

6.3 Chip-to-chip Communication Networks

Transmission line effects become significant when the round-trip propagation delay from the sending chip to the receiving chip is greater than the rise time of the transmitted signal. This condition is almost always met for modern CMOS-based digital systems. When this is the case, reflections occurring on the signal line due to impedance mis-matches between the source, the transmission line, and the load are not 'hidden' in the rising edge of the signal at its source. Rather, these reflections are superimposed on the transmitted signal, causing significant overshoot, undershoot, and system-wide noise. The result is a reduction in the valid-data window of transmitted pulses with respect to the system clock.

This section examines practical transmission line signal networks and leaves detailed discussions of transmission line effects to the wide body of available literature [6.12]. It is useful to recall, however, a few transmission line basics. The characteristic impedance for a transmission line having an inductance per unit length L, a capacitance per unit length C, and a low resistance may be approximated as:

$$Z_0 = \sqrt{\frac{L}{C}}$$

Any impedance discontinuity in a transmission line, from an impedance Z_1 to an impedance Z_2, results in a partial reflection of the propagated signal at the discontinuity. The fraction of reflected signal, or reflection coefficient, is given as:

$$\rho = \frac{Z_2 - Z_1}{Z_2 + Z_1}$$

Finally, the velocity of signal propagation along a low resistance transmission line is:

$$v = \sqrt{LC}$$

where L and C are again per unit length values.

---

3. Section 6.3.2 discusses multiple-load terminated networks relevant to the signaling approaches of both the SLDRAM and RDRAM standards.

6.3.1 Point-to-Point Networks

A point-to-point network consists of a single driver and receiver pair. In systems where the distance between the driver and receiver may be kept short and source and load impedances may be carefully controlled, these networks may be series-terminated, as illustrated in Figure 6.3(b).[4] Series termination is preferable because there is no DC power consumption in the signal network.

> **Rule of thumb:** If the distance from driver to receiver is short enough that the round-trip transmission line delay is less than half the pulse width of the transmitted data, point-to-point series termination may be used.

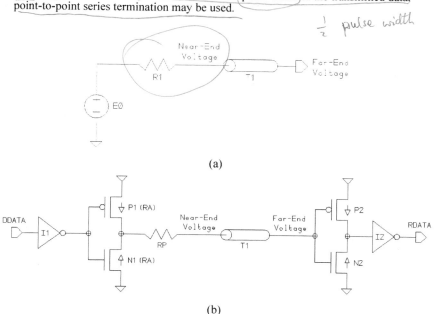

(a)

(b)

FIGURE 6.3 (a) Ideal point-to-point series termination with infinite load. (b) CMOS point-to-point series terminated network.

Ideally, the far end of the series-terminated transmission line is an infinite impedance. This is very nearly the case for CMOS circuits, where the gates of the receiving devices in Figure 6.3(b) are small. The total load capacitance at the far end of the transmission line, including the receiver circuitry, ESD circuitry, and package connec-

---

4. Series termination is also referred to as source termination or source-series termination.

tions, is usually 5 to 10 pF. This capacitance introduces some inherent impedance mismatch in the system.

Figure 6.4(a) illustrates the near-end and far-end voltages resulting from a single data pulse on the ideal series-terminated configuration of Figure 6.3(a) for the case where R1 equals the characteristic impedance of T1 (50 ohms) [6.9]. The near-end voltage plateaus at a potential of $V_{DDQ}/2$ while the transmission line is being charged. During this plateau, the current through the driver remains constant, as illustrated in Figure 6.4(b). The current ramp-rate to reach this plateau, as well as the ramp-rate back to zero DC current once the transmission line is fully charged, determines the amount of simultaneous switching noise introduced by the driver in a non-ideal system. Various methods for controlling this ramp-rate are discussed in Section 6.5.2.

Driver power supply noise can result in impedance mis-matching between resistor R1 and the characteristic impedance of transmission line T1. This mis-matching occurs because R1 is formed by both the switched active resistance, RA, of the CMOS driver output devices and any series passive resistance, RP, built into the driver. The value of RA depends on both the power supply potentials at the sources of devices N1 and P1 of Figure 6.3(b) and the potential at the common drain of these devices. Thus variations in the power supply potential vary RA and cause multiple signal reflections of the transmitted pulse due to impedance mis-matches. Similarly, variations in the near-end voltage also cause RA to vary with time. These variations may caused by signal reflections due to the capacitive transmission line termination or by impedance discontinuities in the signal network represented by transmission line T1. Superimposing the reflections of all these mis-matches can result in significant system noise.

Figure 6.5(a) and Figure 6.5(b) illustrate the effects of mis-matched source and transmission line impedances. Figure 6.5(a) illustrates an under-damped response to a source impedance (30 ohms) which is lower than the transmission line characteristic impedance (50 ohms). This technique is often used to improve the first-incident switching of the far-end voltage, but can result in significant overshoot, a longer settling time due to signal reflections, and increased power consumption if the source impedance is too low. The slow response time of the far end voltage in Figure 6.5(b) results in less power consumption, but is generally not desirable in high performance systems.

6.3.2 Multiple-Load Networks

Multiple-load networks are common in microprocessor applications where, for example, a single driver must supply an address signal to multiple external devices. In such a network, the total capacitance of the receiving devices may be significant and the distance to these devices from the driving device may result in a long round-trip trans-

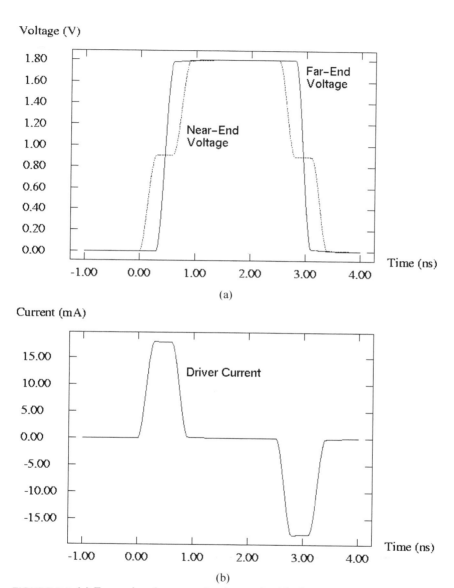

FIGURE 6.4 (a) Far-end and near-end voltages for ideal, series-terminated, 50 ohm driver and 50 ohm transmission line characteristic impedance. (b) Current through driver for this series-terminated configuration.

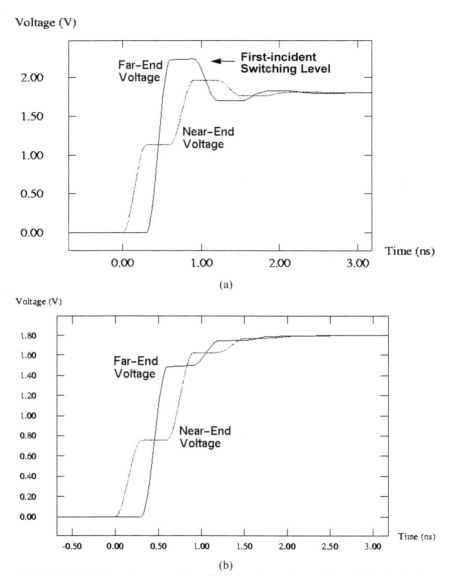

FIGURE 6.5 (a) Overshoot resulting from low-impedance driver (30 ohm) mis-match with 50 ohm transmission line. (b) Slow far-end voltage rise time resulting from high-impedance driver (70 ohm) mis-match with 50 ohm transmission line.

mission line delay. Signal reflections in series-terminated networks result in unacceptable switching delays under these conditions: load termination is used instead.

Figure 6.6 illustrates two ideal load-terminated configurations common to HSTL, STTL, and other small-swing interface protocols. Passive load termination results in reduced amplitude signals, since the current path from a power supply of the driver to the termination voltage produces intermediate DC levels for high and low data. These reduced levels do not result in reduced noise margin when carefully implemented, since impedance-matched terminations result in the absorption of signal reflections and thus a reduction in system noise.

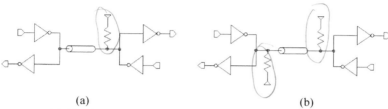

(a)   (b)

FIGURE 6.6 (a) Single resistive termination and (b) double resistive termination for bi-directional networks.

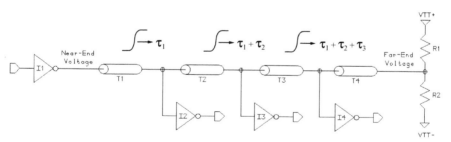

FIGURE 6.7 Typical configuration for terminated high-speed addressing bus. R1 = R2 = 100 ohms. Driver impedance of I1 = 50 ohms.

Impedance matched terminated networks require only a single propagation delay to achieve full switching from one DC level to another at any point along the signal line. Unlike in Figure 6.4, the near-end voltage for a network such as that in Figure 6.7 does not plateau at an intermediate level while charging the entire signal line. Instead, the near-end voltage of Figure 6.7 immediately makes a complete transition, with the near-end rise time depending on the impedance of driver I1 and the impedance of the transmission line. Thus the driver may be switched at a higher frequency than in a series-terminated signal network. In very high frequency communication networks, the driver is switched as soon as possible once the near-end voltage has reached its DC level for the previous output signal polarity. Signals are then allowed to propagate from the source device to multiple receiving devices without charging the entire signal line to a DC state. This technique results in a signal delay from one receiving chip to the next in Figure 6.7: a delay of τ_1 to receiver I2, $\tau_1+\tau_2$ to receiver I3, etc. Data integrity is maintained at each receiving device as long as the system clock propagates in the same manner.

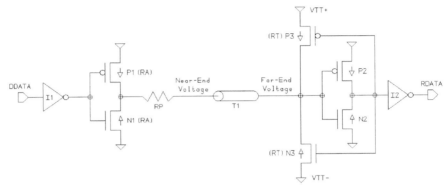

FIGURE 6.8 Self-terminating receiver circuit. Terminated to voltages $V_{TT}+$ and $V_{TT}-$ through the active impedance RT of devices P3 and N3.

Figure 6.8 illustrates active load termination using a self-terminating receiver [6.10]. This configuration represents a compromise in power and performance between series termination and passive termination in multiple-load networks. The output of receiver devices P2 and N2 creates a feedback loop into devices P3 and N3. Together, these four devices form a latch, with P3 and N3 also providing active termination of the transmission line to $V_{TT}+$ and $V_{TT}-$ respectively at the receiver input. $V_{TT}+$ and $V_{TT}-$ are typically the same supply potentials as for devices P2 and N2. The impedance of the driver, RA+RP, must be low enough that first incident switching at the far end of the transmission line causes the receiver-latch to switch.[5] The time required to switch the latch results in one reflection of the transmitted data. A shorter latch switching time results in less power dissipation through devices N3 and P3 as well as a smaller amplitude signal reflection, and thus a shorter settling time of the signal line. Subsequent reflections are significantly reduced through the terminating impedance, RT, of either NFET device N3 or PFET device P3, making the self-terminating

5. See Figure 6.5 for an illustration of first-incident switching.

receiver suitable for multiple-load networks [6.11]. Although this method still requires the entire signal line to reach a DC state before switching the driver, the benefits are the elimination of DC power consumption with a reduction in signal settling time and system noise.

6.4 ESD Protection

Electrostatic discharges subject the I/O pins of a semiconductor device to high-voltage, high-current stresses. These stresses can cause long-term reliability problems or catastrophic failure the in I/O circuitry of a chip. ESD protection circuits provide low resistance paths under high-voltage conditions to dissipate the energy in ESD pulses.

Three standard models for ESD pulses exist: the human body model (HBM), the machine model (MM) and the charged-device model (CDM). HBM pulses represent a discharge of a 1kV pulse over nearly 200 nS, an approximation of a human body contacting a grounded device. MM pulses are described as damped, oscillatory current sources with peak currents of 7 to 16 A at frequencies of 7 to 16 MHz. These pulses represent ESD events which can occur during machine handing. CDM pulses are lower in voltage, but longer in duration than HBM and MM pulses, representing the flow of charge accumulated on-chip to an external ground [6.13].

These ESD pulses can cause irreversible damage such as fused metal wires[6] and thermal breakdown of PFET and NFET devices. Figure 6.9 illustrates drain current versus drain potential for the five operating regions of an NFET device with a positive gate to source potential. This curve extends beyond the region of typical device modeling to show the behavior of the NFET device under high-current conditions.

The linear and saturation regions of Figure 6.9 are the most familiar and best modeled regions of device operation. Avalanche breakdown occurs when substrate currents begin to flow from the P-doped substrate beneath the gate of the NFET to the N+ source diffusion. These substrate currents increase until the parasitic lateral NPN bipolar device formed by the N+ source, P channel, and N+ drain diffusions of the NFET is turned on. The NFET then enters the snapback region. The activation of this bipolar parasitic results in significant current gain and thus an initial voltage drop at the drain. This voltage drop is followed by large increases in drain current for small

---

6. Fused metal wires are open circuits which result from thermal fracture or melting of the conducting material.

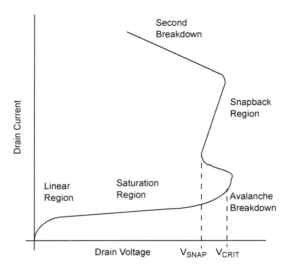

FIGURE 6.9 General drain current vs. drain voltage curve for an NFET with a non-zero gate to source voltage including time-dependent high-current thermal effects.

increases in drain voltage. Thermal breakdown eventually occurs when the high drain current causes localized silicon heating and irreversible damage.

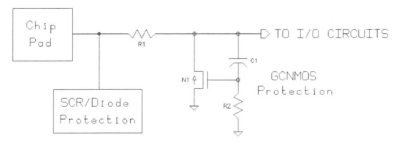

FIGURE 6.10 ESD configuration with SCR and diodes as the primary protection and a resistively isolated GCNMOS device as the secondary protection.

Figure 6.10 illustrates a common ESD protection topology to prevent irreversible damage to CMOS devices. This method uses either silicon controlled rectifiers (SCRs) or diodes (or both) for primary ESD protection and NFET device N1, a gate-coupled NMOS (GCNMOS) device, for secondary protection. This protection circuitry requires significant area to dissipate large currents, adding 1 to 3 pF of capacitance to the chip pad. This circuitry merits careful optimization to minimize capacitive loading of high frequency I/O signals.

During an ESD event, capacitance C1 couples the source of N1 to the gate of the device, causing the NFET to enter the snapback region of operation more rapidly. Once in the snapback region, device N1 directs charge which would otherwise be imposed on the I/O circuitry to ground.

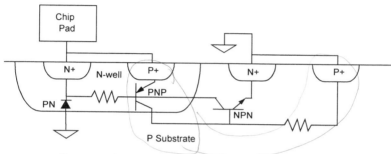

FIGURE 6.11 Cross section of lateral SCR device with PNP, NPN, PN, and resistive elements identified.

The same parasitic bipolar devices which can unintentionally lead to latch-up in CMOS circuitry are used to create SCR devices with high current-gain as ESD protection structures. Figure 6.11 illustrates a cross section of a basic SCR structure. A positive transient at the chip pad forward biases the emitter to base junction of the PNP. The resulting collector current raises the base potential of the NPN, forward-biasing the base to emitter junction. Once the NPN is activated, positive feedback further decreases the PNP base potential (as well as the chip pad potential) and increases the NPN base potential. This positive feedback allows the SCR to dissipate significant current during a positive ESD event [6.14]. For a negative ESD pulse, the illustrated P substrate to N-well diode forward biases to limit the potential at the chip pad.

Figure 6.12 illustrates a double-diode structure which may be used as a primary ESD protection circuit for interfaces where the input signal applied to the chip pad does not normally exceed the internal power supply of the chip. For mixed voltage interfaces, the PN diode from the chip pad to V_{DD} can inadvertently become forward-biased when the chip pad is externally driven to a voltage 0.6 V greater than V_{DD}. To avoid forward biasing the ESD diode, multiple diodes are used in series from the chip pad to V_{DD} [6.15].

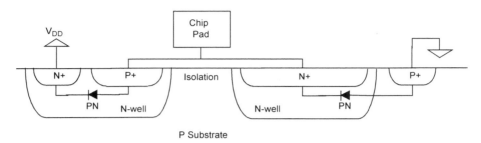

FIGURE 6.12 Cross section of double-diode structure for ESD protection.

6.5 Driver Design Techniques

This section reviews several driver circuits and sub-circuits of general interest for a variety of high-speed applications. The implementation of any of these topologies requires consideration of the access time benefits of a minimized delay through the driver versus the need for low simultaneous switching noise and matched high to low and low to high transitions. Although the driver output stages are illustrated in their fully complementary forms, all of these circuits may also be used as either pull-up or pull-down networks in a terminated environment.

6.5.1 Pre-Drive Generation

The output stage of a CMOS output driver uses PFET and NFET devices which are hundreds to thousands of times wider than most devices used in internal logic circuitry. These wide channel dimensions are necessary to achieve sufficient current sourcing and sinking capability for charging and discharging external loads. The size of these devices, however, requires that buffering circuitry, commonly referred to as pre-drive circuitry, be used to control the turn-on and turn-off characteristics of these output devices. This pre-drive circuitry acts as a buffer between internal registers or logic outputs and the large output devices of the driver.

Figure 6.13 illustrates a basic cascaded output driver topology. An internal DATA signal is buffered through inverter stages I1, I2, and I3 before being applied to the final output devices, PFET device P1 and NFET device N1. This kind of buffering scales inverters I1 through I3 successively by a factor of 3 to 8 [6.9]. Although most pre-drive circuitry uses some kind of cascading, there are significant limitations in using the simple driver of Figure 6.13.

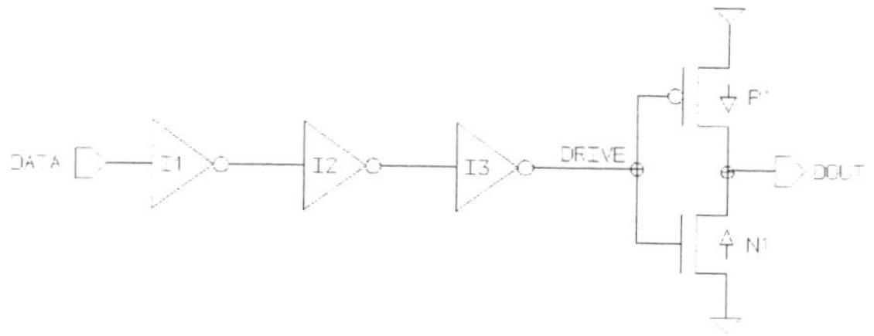

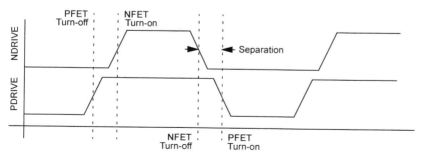

FIGURE 6.13 Cascaded output driver.

The pre-drive topology of Figure 6.13 applies the same drive signal to PFET P1 as to NFET N1. For typical logic circuits, the peak switching current of a CMOS inverter is less than 5 mA and short in duration. When referring specifically to driver currents, the portion of switching current which flows directly from the positive supply of the driver to the negative supply is also known as shoot-through current. Shoot-through currents can exceed 50 mA per driver; the duration of these currents increases with increasing rise and fall times of the DRIVE signal applied to the output stage. The amount of current sourced to the load when both NFET device N1 and PFET device P1 are active depends on the termination configuration of the external load. It has been shown that for series terminated loads, nearly all of this current appears as shoot-through current [6.16]. Thus the turn-on current characteristics of the series ter-minated driver are dominated by the switching current, making simultaneous switch noise difficult to limit when the NFET and PFET devices are controlled with the same signal.

The pre-drive circuit of Figure 6.13 also does not allow the driver to be placed in a high impedance state. Separating the PFET and NFET driving signals allows both devices to be turned off simultaneously for bi-directional I/O ports.

FIGURE 6.14 Illustration of separated PDRIVE and NDRIVE signals.

Two common approaches to separating the PFET output device drive signal from the NFET output device drive signal are illustrated in Figure 6.15 [6.16][6.22]. Both of these topologies allow drive signal buffering to be built into the pre-drive circuitry. These circuits may be tuned to increase or decrease the separation in time between the PFET drive and the NFET drive signals illustrated in Figure 6.14. Too much separation between these signals results in unnecessary delay through the driver and, at high frequencies, less signal development at the load. Too little separation between these signals results in avoidable shoot-through currents.

> **Rule of thumb:** The separation between the NDRIVE and PDRIVE signals should not exceed the 10% to 90% rise/fall time of these signals unless additional turn-on and turn-off staggering occurs in the output stage (see Section 6.5.2).

Figure 6.15(a) uses NAND and NOR gates to drive devices P1 and N1, respectively. These logic gates may be skewed such that the rise time of ND1 is faster than its fall time and the fall time of NR1 is faster than its rise. Skewing the gate outputs in this manner results in non-overlapping PFET drive and NFET drive signals. A low enable signal forces node NDRIVE low and node PDRIVE high, placing the driver in the high impedance state [6.16].

The topology of Figure 6.15(b) combines the PFET and NFET drive generation into a single circuit [6.22]. When the ENABLE input signal is high, the pass gate formed by devices N3 and P3 is active. This pass gate causes node PDRIVE to fall slightly slower than it rises, since this node is pulled up through PFET device P1, but pulled down through NFET device N1 and the pass gate in series. Likewise, node NDRIVE falls slightly faster than it rises, being pulled down through only NFET device N1 and pulled up through PFET device P1 and the pass gate. When the ENABLE input is low, the pass gate is disabled and PFET device P2 and NFET device N2 are active, forcing node PDRIVE high and node NDRIVE low. Under these conditions, both PFET device P4 and NFET device N4 are off: the driver is then in the high impedance state.

These and other similar pre-drive circuits may be used in conjunction with any of the output stages presented in the following sections.

6.5.2 Output Stages and Driver Impedance

Controlling the amount of current which the output stage delivers and the rate at which it supplies this current to the external load is critical for minimizing simultaneous switching noise. Simultaneous switching noise is proportional to the time-rate change of current through the driver, dI/dt, and the number of drivers switching. Thus for wide output buses, small decreases in the current ramp rate through a single driver

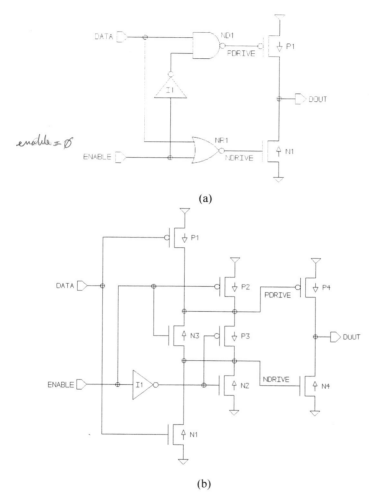

(a)

(b)

FIGURE 6.15 (a) Pre-drive circuitry using NAND/NOR topology for separate PFET and NFET output control with high impedance capability. (b) Pre-drive circuitry using pass gate topology to implement the same function.

can significantly reduce the self-induced simultaneous switching noise of the outputs and thus enhance system-wide performance.

Three basic approaches exist for controlling output impedance and the corresponding current delivery through the output devices:

1. Fixed Impedance
2. Variable Impedance
3. Current Limiting

The fixed impedance method assumes a near-end voltage which remains essentially constant during switching. By knowing the voltage conditions under which the device operates during switching, the resistance of the output device may be tuned to a desired value. The combined driver resistance and load impedance then determines the maximum current delivery through the driver for a given power supply voltage. The effects of near-end voltage variations from reflections on the bus may be minimized by placing a large resistor in series with the driving FET so that the largest variations occur across the passive resistor instead of the active device. This method results in very wide driver devices, particularly when a low output resistance is necessary. Another method for limiting variations in the driver impedance with near-end voltage variations is to use NFET and PFET devices in parallel as both pull-up and pull-down devices [6.17]. This method produces good results, but requires generating four driving signals: two complementary signals for each pair of pull-up and pull-down devices. Accurately generating all four signals at high frequencies can be challenging.

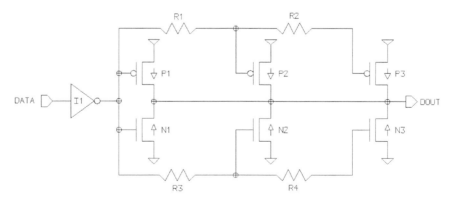

FIGURE 6.16 Passive, three-stage dI/dt control output driver with separate PFET and NFET output control signals.

Figure 6.16 illustrates a method for reducing the current ramp rate through a fixed-impedance driver [6.18]. Resistor R1 causes an RC delay between the gate signal of PFET device P1 and that of PFET device P2. Likewise, R2 places a similar delay between the gate signals of P2 and P3. The NFET controls are similarly delayed. The resulting delays stagger the switching of the PFET and NFET devices.

> **Rule of thumb:** The benefits of turn-on and turn-off staggering depend on the voltage and current requirements of the output driver, but are generally not significant beyond three stages of staggering [6.16].

The staggering of Figure 6.16 may be implemented with active delay devices instead of passive delay devices [6.19][6.17][6.23]. Replacing resistors R1-R3 with buffers or pass gates has the advantage of making the rise times of the drive signals applied to PFET devices P1-P3 and NFET devices N1-N3 more uniform, but can also result in too much staggering between stages in high-speed applications.

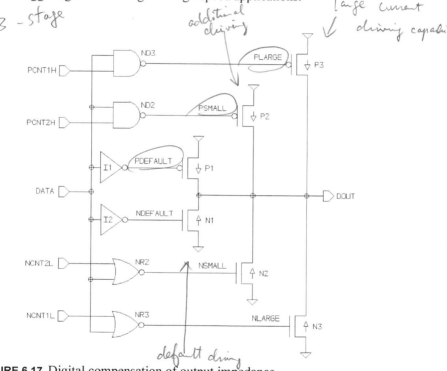

FIGURE 6.17 Digital compensation of output impedance.

Figure 6.17 illustrates a common approach for variable impedance tuning of the output stage of the driver [6.21][6.22]. The signals PLARGE, PSMALL, and PDEFAULT control which combination of PFET devices P1 through P3 are active when the driver is passing a high voltage value to the DOUT node. Signals NLARGE, NSMALL, and NDEFAULT provide similar control for the NFET output devices.

The more FETs which are active when the DOUT signal is driven high or low, the lower the impedance of the driver. The 'n' subdivisions of the total output width are often binary-weighted to allow (2^n - 1) total-width combinations. Default devices P1 and N1 are always switched during driver operation and determine the highest possible impedance at which the driver may be operated.

The control signals PCNT1H, PCNT2H (high-true for PFET output device activation), NCNT1L, and NCNT2L (low-true for NFET output device activation) of Figure 6.17 can be generated using a feedback control system. This feedback control compares the result of voltage division between a set of PFET or NFET devices identical to those used in the output driver to either an external resistance or an on-chip resistance [6.22]. A careful implementation of this feedback control allows the impedance of the driver to be matched to the reference resistance independent of process, temperature, and voltage variations. The impedance range of the driver is limited by the total width of the PFET or NFET output devices in parallel and the variation in device length across process. The impedance discretization is limited by the number of impedance control signals and the smallest device width which can be accurately reproduced in a given technology.

The staggering technique of Figure 6.16 may be applied to the driver of Figure 6.17. Subdividing P1-P3 and N1-N3 each into three staggered devices retains the same output dimensions with a total of 18 distinct devices. Another approach is to stagger the switching of the default devices with respect to the switching of the programmable devices. Staggering the switching of the programmable devices with respect to each other may not be desirable, since the delay through the driver would vary with each combination of programmable devices.

The same feedback control signals applied to the driver of Figure 6.17 may be used to digitally adjust the capacitive loading of the pre-drive circuitry. As discussed in Section 6.5.1, the rise times of pre-drive signals PLARGE, PSMALL, PDEFAULT, and the corresponding NFET controls affect the switching characteristics of the output devices. Increases or decreases in the rise times of these signals due to process variations in the output device channel lengths correspond to increases or decreases, respectively, in the simultaneous switching noise of the drivers. Figure 6.18 illustrates the digital compensation of the capacitive loading of the pre-drive signals [6.22]. As programmable devices in the driver are made inactive to compensate for shorter channel lengths (i.e. more current per width of device), additional capacitance is switched into the pre-drive signal path to the output devices. This method is only applicable when the digital impedance control of the driver is implemented to achieve a single, pre-established impedance value. Such capacitive compensation is not practical when a range of programmable impedances is desired.

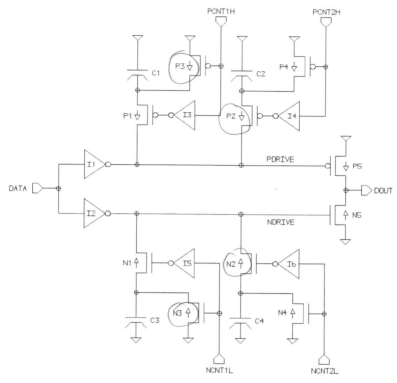

FIGURE 6.18 Digital compensation of pre-drive capacitive load.

Figure 6.19 illustrates a current-limiting output driver. In this topology, the PBIAS and NBIAS voltages applied to device P2 and N2 respectively limit the maximum current through these devices. These biases may be generated using current mirroring techniques and on-chip reference circuits [6.20]. The NFET pull-down portion of this circuit has been shown to achieve high frequency performance in a terminated environment [6.24]. Turn-on and turn-off staggering may be implemented more easily in this topology than in the variable impedance topology of Figure 6.17 by merely staggering PFET device P3 and NFET device N3 as in Figure 6.16.

Devices N3 and P3 may also be eliminated from the circuit of Figure 6.19. The PBIAS node then switches between V_{DD} and a DC bias level, while NBIAS switches between ground and a DC bias level. A controlled current ramp rate through the driver may still be achieved by dynamically varying the current through J1, thus varying the DC bias levels of the NBIAS and PBIAS nodes.

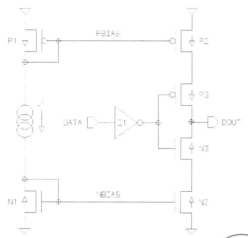

FIGURE 6.19 Current limiting output driver using current mirroring technique.

6.5.3 Output Voltage Translation

Two distinct problems must be addressed for short-channel CMOS output drivers which operate on low internal power supply voltages: driving outputs to voltage levels greater than the internal supply voltage and preventing leakage current in the high impedance state of the driver when a bus connecting several devices is externally driven to a voltage greater than the on-chip output driver supply.

Low V_{DD} to High V_{DDQ} Translation

Figure 6.20 illustrates a cross-coupled PFET technique for voltage translation from low internal levels to higher external levels [6.25]. The DATAC input to this circuit transitions between ground and V_{DD}. The PDRIVE signal transitions between ground and V_{DDQ} and is applied directly to the PFET output device of the driver. When DATAC is low, NFET device N1 is off and NFET device N2 is on. Node PDRIVE is pulled low through NFET device N4, turning PFET device P1 on to pass a full V_{DDQ} level to the gate of PFET device P2. P2 is then completely off. When DATAC transitions to V_{DD}, N2 is off and N1 is on, causing node PDRIVE to float until the gate of P2 is pulled low, forcing PDRIVE to V_{DDQ}. NFET devices N3 and N4 provide dielectric stress and hot-electron protection in the pull-down paths. When the difference between V_{DD} and V_{DDQ} is small, these two devices may be eliminated [6.26]. For cases where $V_{DDQ} \gg V_{DD}$, similar cross-coupled PFET techniques may be used to switch PDRIVE between V_{DDQ} and an intermediate DC level. Switching PDRIVE to an intermediate DC level eliminates reliability concerns in the PFET pull-up path of this topology.

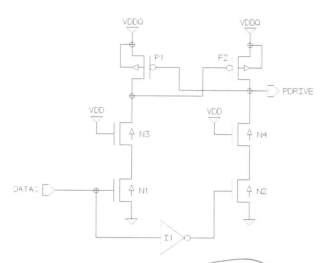

FIGURE 6.20 Output voltage translation using cross-coupled PFET devices.

Interfacing to voltage levels where dielectric stress and hot-electron effects are a concern requires stacking the output devices. Dielectric stress is a concern when the maximum output level plus the AC clamping voltage of the ESD protection causes either the gate to drain or gate to source potentials of the output devices to exceed the technology limit. When interfacing to levels very close to the technology limit, the maximum potential across the gate dielectric can often be adjusted for the duty cycle of the output driver, since significant stress levels occur only during brief signal transients.[7] Hot-electron effects must be considered when the maximum output level exceeds the junction potential limit of the technology.

> **Rule of thumb:** Output device stacking is generally necessary for sub-quarter micron technologies when interfacing to levels of 2 V or greater in single-oxide thickness technologies.

Figure 6.21 illustrates a stacked-device driver output stage. Voltage translation, such as that provided in Figure 6.20, is required to generate a PDRIVE signal which transi-

---

7. See Figure 6.2(b) or Figure 6.5(a).

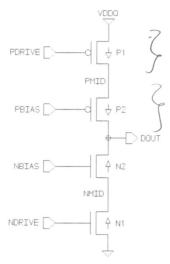

FIGURE 6.21 Stacked output driver topology for limiting dielectric stress and hot-electron effects.

tions between ground and V_{DDQ}. NDRIVE, however, may transition between internal supply levels.

The output stack of Figure 6.21 is similar to that of Figure 6.19, except that the bias voltages are now applied to the inner two devices, P2 and N2, while the outer devices, P1 and N1, are switched. PBIAS and NBIAS are once again DC voltage levels. The PBIAS voltage is set low enough to provide good current drive through PFET device P2 and prevent the difference between the PBIAS and DOUT from exceeding the dielectric stress limit of the technology during a negative excursion of DOUT. At the same time, the PBIAS voltage is set high enough to also prevent the difference between the PBIAS and PMID nodes from exceeding the technology limit. Similar constraints apply for selecting the voltage applied to the NBIAS node.

When DOUT in Figure 6.21 transitions from ground to V_{DDQ}, P1 and P2 are both active and provide voltage division between V_{DDQ} and DOUT. This voltage division prevents the potentials between V_{DDQ} and PMID and between PMID and DOUT from exceeding the forward-bias junction potential limit of the technology. Devices N1 and N2 provide similar voltage division when N1 is active, causing DOUT to transition from high to low. In cases where $V_{DDQ} \gg V_{DD}$, PDRIVE may be switched between V_{DDQ} and a bias voltage greater than ground to prevent excessive dielectric stress across PFET device P1 when driving DOUT to a high voltage potential.

Stacked driver topologies are necessary to prevent reliability concerns, but are limited in their output frequency. The impedance of these drivers can be difficult to adjust due to capacitive coupling through devices P2 and N2 onto the PBIAS and NBIAS voltages, respectively. As PFET device P1 turns on, the source of P2 can quickly charge to an intermediate voltage, causing the PBIAS node to couple up capacitively and increase the total impedance of the pull-up path through devices P1 and P2. NBIAS is similarly coupled low capacitively in a high to low transition.

Another limitation to high frequency performance in a stacked driver is the significant increase in output capacitance. The device widths must be doubled relative to the widths of an unstacked driver to achieve the same output impedance. This increased width doubles the capacitance on node DOUT. The capacitance is further increased because P2 and N2 are always active, causing the output capacitance to include not only the source capacitance of P2 and N2, but also the gate and diffusion capacitances coupled to nodes PMID and NMID.

Multi-Level Bus Drivers

In mixed voltage interfaces where several devices, fabricated in various technologies, must communicate using the same data bus, the off-chip driver must be able to maintain a high-impedance state independent of the voltage applied to its output node. NFET output devices are commonly not a concern, as long as all devices on the bus drive to the same ground or negative supply for a low output signal. It is possible, however, for high output levels to span 1.8 V or lower to 5 V on the same signal bus. When node DOUT of Figure 6.21 rises two threshold voltage levels above V_{DDQ}, both PFET devices P1 and P2 will be on, conducting current from the data bus to the driver power supply. In addition, the biasing of the N-well implant for PFET device P2 to V_{DDQ} creates a problem when node DOUT rises approximately 0.6 V above V_{DDQ}. At this potential, the P-N junction formed by the PFET P+ source diffusion connected to DOUT and N-well implant of device P2 becomes forward biased. This forward-biased diode conducts additional current from the data bus to the power supply of the driver.

A floating N-well driver topology prevents the PFET output device from turning on by dynamically varying the N-well bias voltage. Figure 6.22 illustrates a typical floating N-well topology [6.27]. Several variations of this fundamental topology exist [6.28][6.29]. When the ENABLE input signal is high, the driver output devices operate similarly to those of Figure 6.21, except that the N-well implant common to PFET devices P1, P2, P3, and P4 is charged to V_{DDQ} through P2 when output node DOUT is low. When node DOUT is high, the N-well potential has no active connection and thus 'floats' at its pre-charged V_{DDQ} potential. When ENABLE transitions low, however, the driver goes into a high impedance mode where PDRIVE is driven to V_{DDQ},

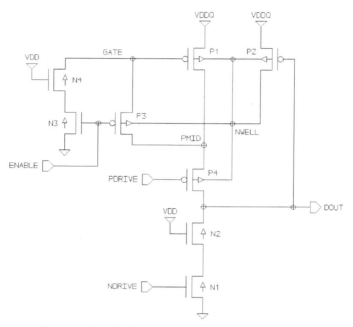

FIGURE 6.22 Floating N-well driver for multi-level bus networks.

NDRIVE is grounded, and PFET device P3 is on. Device P3 shorts the gate of P1 to its source, node GATE to node PMID respectively. When DOUT is then externally driven to a high voltage level, PFET device P1 remains off, even if DOUT exceeds V_{DDQ} by a threshold voltage, since the GATE and PMID nodes are at the same potential. As DOUT forward biases the P2 source to N-well diode, the N-well charges to approximately 0.6 V below the DOUT potential without leaking current back into the driver supply voltage.

6.6 Receiver Design Techniques

As with driver topologies, many receiver topologies are common. These circuits do not generally emphasize self-induced switching noise. Receiver input device dimensions tend to be 10% or less the size of driver output dimensions, making their simultaneous switching a relatively small source of noise. In addition, these circuits are often powered by the more stable and less inductive common power supply of the chip, further reducing their self-induced noise. Receivers can be very sensitive, however, to V_{DD}, ground, input signal, or reference signal noise. This sensitivity can degrade the matching of low to high versus high to low receiver output transitions. Mis-matched receiver output transitions result in a loss of either set-up or hold time at the input latch, requiring that the entire chip operation be slowed to recover these important timings.

This section examines fundamental receiver topologies which continue to be practical as well as more recently developed circuit techniques particularly suited to high-speed applications. As with the driver circuits of Section 6.5, a minimized delay through the receiver must be balanced with the need for matched receiver output transitions as well as the need for input voltage translation.

6.6.1 Receiver Amplification Stage

The two most common CMOS receiver implementations are single-ended and differential, corresponding to the output modes mentioned in Section 6.1.1. In its most basic form, a single-ended receiver is a push-pull amplifier which has been tuned to a specific switch point. This switch point is typically the midpoint between the $V_{IH}(min)$ and $V_{IL}(max)$ values of the input. To achieve switch points which are significantly greater than or less than $V_{DD}/2$, stacked NFET devices may be used in the pull-down path (see Figure 6.23(b)) or stacked PFET devices in the pull-up path, respectively. NFET stacking is common when the input voltage swings exceed the internal power supply.

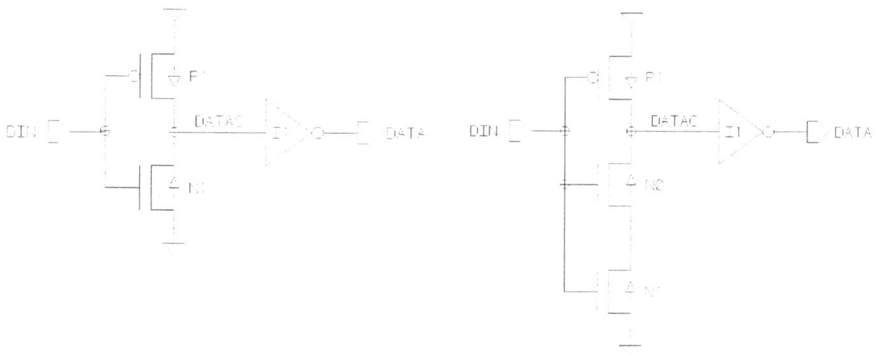

(a)   (b)

FIGURE 6.23 (a) Single-ended inverter-receiver. (b) Single-ended receiver with raised switch point.

Significant timing changes through the single-ended receiver occur when the input and either of the receiver's power supplies are subject to differential noise. Consider an increase in the ground potential of Figure 6.23(a) relative to the off-chip ground supply to which the DIN signal is referenced. DIN must reach a higher potential than normal in order to exceed the threshold voltage of NFET device N1 and cause node DATAC to begin to transition. No such delay occurs if DIN transitions low, assuming V_{DD} is not subject to the same noise as ground. Thus any low to high transition will be slightly delayed relative to a high to low transition, causing timing inaccuracies among any synchronous, latched inputs.

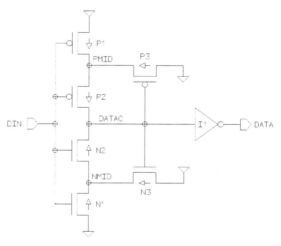

FIGURE 6.24 Single-ended inverter-receiver with hysteresis.

Hysteresis may be added to the CMOS inverter-receiver topology as illustrated in Figure 6.24 [6.30][6.31]. When node DIN is low and DATAC high, NFET device N3 is on, precharging node NMID to one threshold voltage below V_{DD}. On a low to high transition of DIN, node DATA must transition high enough to allow NFET device N1 to pull down more strongly than N3 pulls up. The potential at DIN must also exceed the threshold voltages of both N1 and N2. The effect of this hysteresis is symmetric for a high to low transition of DIN, providing noise immunity in both cases once DATAC has switched. Hysteresis is beneficial in noisy systems, but often abandoned in high performance circuits to improve the matching of high to low versus low to high timings and to reduce receiver delay.

A differential input stage amplifies an input signal relative to either the common-mode voltage of its complement, or relative to a reference voltage having the same

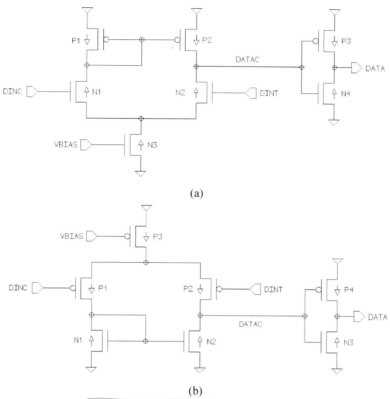

(a)

(b)

FIGURE 6.25 (a) NFET differential amplifier receiver. (b) PFET differential amplifier receiver. In (a), VBIAS is often connected to the gate of PFET device P1. Similarly, in (b), VBIAS is often connected to the gate of NFET device N1.

common mode potential as the input signal. In the case where a reference level must be provided from off-chip to several receivers, particular care must be taken to isolate this signal capacitively from any on-chip signals having significant AC noise.

Figure 6.25(a) illustrates an NFET implementation of a differential amplifier stage: Figure 6.25(b) illustrates a PFET implementation [6.32]. This method significantly reduces the sensitivity of the receiver to supply and input noise compared to a single-ended receiver circuit. The NFET implementation is most sensitive to ground noise. The gain from DINT to node DATAC in Figure 6.25(a) decreases with a decrease in the current through NFET device N3 caused by an increase in the source voltage of N3 relative to the analog bias level VBIAS. Likewise, the gain from DINT to node

DATAC in Figure 6.25(b) decreases with a decrease in the source voltage of PFET device P3 relative to node VBIAS.

> **Rule of thumb:** In applications where the drivers and receivers share a common ground supply, the PFET differential amplifier is preferable. If the receivers and drivers share a V_{DD} supply, the NFET topology is more desirable.

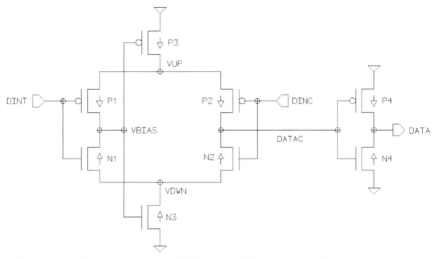

FIGURE 6.26 Complementary self-biasing differential amplifier receiver.

Figure 6.26 illustrates a complementary, self-biasing differential amplifier [6.34]. This topology combines the NFET and PFET differential amplifiers to provide improved noise immunity with respect to both the positive and negative supply voltages of the receiver. Noise injected through ground increases the potential at node VDWN, causing VBIAS to increase momentarily until the potential at node VDWN is reduced by negative feedback through NFET device N3. This amplifier has the added benefit of providing an output swing at node DATAC which is more nearly rail-to-rail, compared to the previous NFET and PFET differential amplifier outputs [6.34]. One limitation of this topology is that devices P3 and N3 must both be kept in the linear region of operation to act as nearly-ideal current sources. The common-mode input voltage range must be limited to ensure this operation. Several alternative topologies have been demonstrated for wide-range common-mode inputs [6.34].

All of the previous amplification stages require a single transition of the input about the amplifier switch point to guarantee set-up and hold data windows for latching the output of the receiver. Multiple excursions of the input through the switching point, as

'illustrated in Figure 6.2(b), could result in the improper evaluation of subsequent circuitry. Figure 6.27 illustrates the basic elements of a pre-amplification stage for a current integrating receiver [6.33]. In this implementation, switches S2 and S4 are closed
and then opened to discharge capacitors C1 and C2. Either switch S1 or switch S3
then close to steer current to either C1 or C2 respectively. If the difference between an
input pair is positive, S1 is closed and S3 is open. If the difference is negative, S1 is
open and S3 is closed. The difference between the input signals is not required to
remain either positive or negative. The difference must merely retain one polarity
long enough during an evaluation period for ΔV, the difference in potential between
the positive nodes of capacitors C1 and C2, to be detected as positive or negative
[6.33].

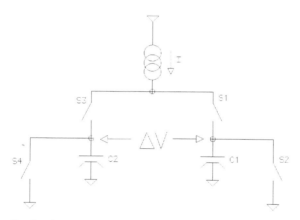

FIGURE 6.27 Basic elements of pre-amplification stage for a current-integrating
receiver.

Figure 6.28 illustrates a circuit implementation of the pre-amplification stage of
Figure 6.27 [6.33]. NFET devices N1 and N2 discharge nodes OUTH and OUTL
respectively for resetting prior to integration. PFET devices P2 and P3 then steer current from PFET device P1 to the combined drain capacitances of either devices P2,
P5, and N1 or devices P3, P6, and N2, respectively. Current integration can result in
cumulative voltage offset errors: these cumulative errors result from charge injection
caused by capacitive coupling of DINC and DINT into nodes OUTH and OUTL
through devices P2 and P3, respectively. Such charge injection causes integration
error when resulting from differential noise on DINT and DINC. Devices P4-P7 limit
this cumulative error by cross-coupling DOUTC to node OUTL and DINT to node
OUTH. The current through PFET device P4 is mirrored as a fraction of the current

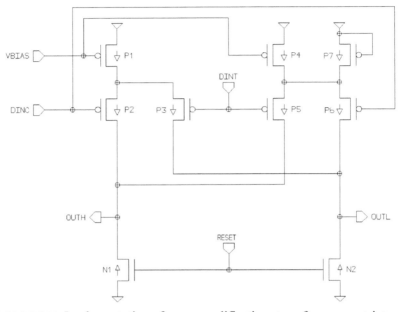

FIGURE 6.28 Implementation of a pre-amplification stage for a current-integrating receiver.

through PFET device P1. This current causes the DC operating points of devices P4-P6 to be comparable to those of devices P1-P3. PFET device P7 ensures that the total width of devices P4 and P7 equals the width of device P1. A positive difference between nodes OUTH and OUTL at the end of an evaluation period indicates a high input to the receiver. This difference must be sampled and then translated to CMOS levels by any of the previously described amplification methods [6.33].

6.6.2 Input Voltage Translation

Figure 6.29 illustrates one basic approach to limiting the input voltage levels imposed on a receiver circuit using a pass-protect topology [cite prior art from John's patent]. NFET pass device N1 limits the up-level voltage at the input to the receiver circuit to a threshold voltage below V_{DD}. This simple configuration is sufficient for wide input voltage swings which must be limited to a smaller internal supply voltage to prevent dielectric stress on the input devices of the receiver circuit. This method does not, however, provide a full V_{DD} level to the receiver, resulting in unnecessary DC power consumption in single-ended receivers. Furthermore, the receiver is not protected from input transient below ground, which may result in significant dielectric stress for any PFET devices with sources connected to the internal V_{DD} supply and gates connected to the receiver input signal.

FIGURE 6.29 NFET pass device receiver input protection.

Figure 6.30 illustrates a pass-protect circuit which provides a higher up-level input to the receiver circuit and also provides undershoot transient protection [6.35]. NFET pass device N1 has a wide channel, causing node NGATE to capacitively couple to a voltage higher than V_{DD} when DIN transitions from low to high. This coupling occurs through the gate of N1 because node NGATE is partially isolated from the V_{DD} supply by PFET device P1. Thus node DINN achieves a higher up-level potential. During a negative input excursion, NFET device N2 turns on once node DINN is a threshold voltage below ground. Resistor R1 is must be large enough that the resulting current through P1, N2, N1, and R1 causes voltage division which prevents node DINN from transitioning low enough to exceed the dielectric stress limit of the receiver input devices. Resistor R1 may be implemented as either a metal or a diffusion resistance.

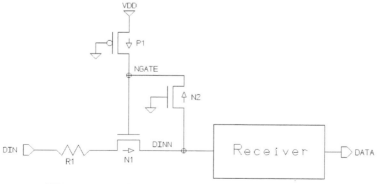

FIGURE 6.30 NFET pass device receiver input protection with resistive element for isolation from undershoot voltages.

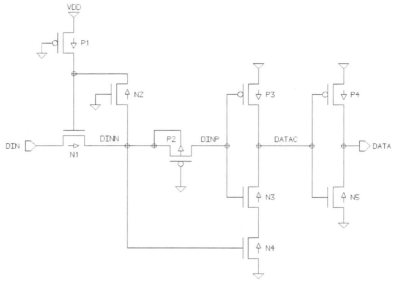

FIGURE 6.31 NFET and PFET devices connected in series with single-ended receiver for combined overshoot and undershoot protection.

Figure 6.31 illustrates a series NFET-PFET pass-protect topology [6.35]. Devices P1, N2, and N1 perform functions identical to those of Figure 6.30. In this implementation, however, a series resistance is no longer necessary. Instead, PFET device P2 prevents node DINR from transitioning below ground. The source connected N-well of PFET device P2 allows node DINP to reach the up-level potential passed by NFET device N1. As this N-well discharges during a low to high transition of DIN, PFET device P2 eventually cuts off when node DINP is a threshold voltage above ground. PFET device P3 is thus protected from undershoot transients. Connecting the gate of NFET device N4 to node DINN ensures zero DC current dissipation through the receiver. NFET device N4 will experience significant dielectric stresses only in extremely negative excursions of the input signal.

6.7 Summary

I/O circuit design requires careful system-level modeling. As clocking frequencies increase with evolving technologies, modeling system-level parasitics with sufficient accuracy for valid design verification becomes critical. Hardware verification of system timings and I/O performance also increases in complexity and difficulty. At the same time, technology advances keep interface levels and system requirements in constant flux. The circuit techniques reviewed in this chapter provide a 'scrapbook' of common approaches and useful circuit topologies for addressing these and other challenges of I/O design.

REFERENCES

[6.1]   "Stub Series Terminated Logic for 3.3 Volts (SSTL_3)," *JEDEC Standard*, EIA/JESD8-8, August 1996.

[6.2]   "High Speed Transceiver Logic (HSTL) A 1.5 V Output Buffer Supply Voltage Based Interface Standard for Digital Integrated Circuits," *JEDEC Standard*, EIA/JESD8-6, August 1995.

[6.3]   "Gunning Transceiver Logic (GTL) Low-Level, High-Speed Interface Standard for Digital Integrated Circuits," *JEDEC Standard*, JESD8-3, November 1993.

[6.4]   "2.5V +/- 0.2V (Normal Range) and 1.8V to 2.7 (Wide Range) Power Supply Voltage and Interface Standard for Nonterminated Digital Integrated Circuit," *JEDEC Standard*, EIA/JESD8-5, October 1995.

[6.5]   "1.8 +/- 0.15 (Normal Range) and 1.2-1.95 (Wide Range) Power Supply Voltage and Interface Standard for Non-terminated Digital Integrated Circuit," *JEDEC Standard*, EIA/JESD8-7, February 1997.

[6.6]   "Standard for Operating Voltages and Interface Levels for Low Voltage Emitter-Coupled Logic (ECL) Integrated Circuits," *JEDEC Standard*, JESD8-2, March 1993.

[6.7]   P Gillingham, "SLDRAM Architectural and Functional Overview," *SLDRAM Consortium*, August 1997.

[6.8]   R. Crisp, "Direct Rambus Technology: The New Main Memory Standard," *IEEE Micro*, November/December 1997.

[6.9]   T. F. Knight Jr., A Krymm, "A Self-Terminating Low-Voltage Swing CMOS Output Driver," *IEEE Journal of Solid-State Circuits*, Vol. 23, No. 2, April 1988, pp. 457-464.

[6.10] F. Anatol, "Zero Power Transmission Line Terminator," *U.S. Patent #5227677*, July 13, 1993.

[6.11] M. Dolle, "A Dynamic Line-Termination Circuit for Multireceiver Nets," *IEEE Journal of Solid-State Circuits*, Vol. 28, No. 12, December 1993, pp. 1370-1373.

[6.12] J. E. Buchanan, "Signal and Power Integrity in Digital Systems," McGraw-Hill, 1996, pp. 153-171.

[6.13] A. Amerasekera, C. Duvvury, "ESD in Silicon Integrated Circuits," John Wiley & Sons, 1995, pp. 9-21.

[6.14] A. Amerasekera, C. Duvvury, "ESD in Silicon Integrated Circuits," John Wiley & Sons, 1995, pp. 73-83.

[6.15] S. Voldman, "ESD Protection in a Mixed Voltage Interface and Multi-Rail Disconnected Power Grid Environment in 0.5- and 0.25-μm Channel Length CMOS Technologies," *Proc. 1994 EOS/ESD Symposium*, pp.125-134.

[6.16] R. Senthinathan, J. L. Prince, "Application Specific CMOS Output Driver Circuit Design Techniques to Reduce Simultaneous Switching Noise," *IEEE Journal of Solid-State Circuits*, Vol. 28, No. 12. December 1993, pp. 1383-1388.

[6.17] H. A. Jungert et al., "Push-Pull Output Stage of Integrated Circuit Providing Enhanced High-Low Voltage Level Signal Transition with Ground Bounce Control," *U.S. Patent #5036232*, July 30, 1991.

[6.18] M. W. Knecht et al., "Output Buffer with Ground Bounce Control," *U.S. Patent #4959565*, September 25, 1990.

[6.19] K Naghshineh, "CMOS Output Buffer Circuit with Improved Ground Bounce", *U.S. Patent #5124579*, June 23, 1992.

[6.20] F. D. Ferraiolo et al., "Current Reference Circuit," *U.S. Patent #5635869*, June 3, 1997.

[6.21]   T. J. Gabara et al., "Digitally Adjustable Resistors in CMOS for High-Performance Applications," *IEEE Journal of Solid-State Circuits*, Vol. 27, No. 8, August 1992, pp. 1176-1185.

[6.22]   T. J. Gabara et al., "Forming Damped LRC Parasitic Circuits in Simultaneously Switched CMOS Output Buffers," *IEEE Journal of Solid-State Circuits*, Vol. 32, No. 3, March 1997, pp. 407-417.

[6.23]   C. S. Choy et al., "A Feedback Control Circuit Design Technique to Suppress Power Noise in High Speed Output Driver," *1995 IEEE Symp. on Circuits and Systems*, pp. 307-310.

[6.24]   J. Navarro et al., "A 1.4Gbit/s CMOS Driver for 50ohm ECL Systems," 1997 *Proc. Great Lakes Symposium on VLSI*, pp. 14-18.

[6.25]   S. H. Dhong et al., "A low-noise TTL-compatible CMOS off-chip driver circuit," *IBM Journal of Research and Development*, Vol. 39, No. 1/2, January/March 1995, pp. 105-112.

[6.26]   B. Keeth et al., "Low-to-high Voltage CMOS Driver Circuit for Driving Capacitive loads," *U.S. Patent #5670905*, September 23, 1997.

[6.27]   R. D. Adams et al., "CMOS Off-Chip Driver Circuits," *U.S. Patent #4782250*, November 1, 1988.

[6.28]   B. Heim et al., "CMOS Output Driver Which Can Tolerate an Output Voltage Greater Than the Supply voltage Without Latchup or Increased Leakage Current," *U.S. Patent #5451889*, September 19, 1995.

[6.29]   T. Nguyen et al., "3.3 Volt CMOS Tri-state Driver Circuit Capable of Driving Common 5 Volt Line," *U.S. Patent #5467031*, November 14, 1995.

[6.30]   H. Hanafi et al., "Design and Characterization of a CMOS Off-Chip Driver/Receiver with Reduced Power-Supply Disturbance," *IEEE Journal of Solid-State Circuits*, Vol. 27, No. 5, May 1992.

[6.31]   Q Mahmood, "Input Buffer Circuit Including an Input Level Translator with Sleep Function," *U.S. Patent #5602496*, February, 1997.

[6.32]   R. L. Geiger et al., "VLSI Design Techniques for Analog and Digital Circuits," Mc-Graw Hill, Inc., 1990, pp. 431-453.

[6.33]   S. Sidiropoulos, M Horowitz, "A 700-Mb/s/pin CMOS Signaling Interface Using Current Integrating Receivers," *IEEE Journal of Solid-State Circuits*, Vol. 32, No. 5, May 1997, pp. 681-690.

[6.34]   M Bazes, "Two Novel Fully Complementary Self-Biased CMOS Differential Amplifiers," *IEEE Journal of Solid-State Circuits*, Vol. 26, No. 2, February 1991, pp. 165-168.

[6.35]   R. D. Adams et al., "Dynamic Dielectric Protection Circuit for a Receiver," *U.S. Patent Pending*, May 13, 1997.

CHAPTER 7

Clocking Styles

7.1 Introduction

The clocking style selected for a given design is dependent on the logic styles to be supported in the design as well as the desired/prescribed latch structure(s). Consequently, the choice of a logic/latch/clocking style is a chicken/egg/nest problem: which comes first?

It is extremely important to note that the overall clocking style for the chip must be decided very early in the design cycle, before actual implementation has begun. This permits the designers to account for the clock skew and routing criteria in their designs as the chip takes on its proper form. Furthermore, it is equally important that the decision on the clocking style and distribution system never be re-opened after the design has begun, save minor tweaking to the skew tolerances, etc., unless absolutely necessary. That is, the clocking style is as key a piece of the foundation of the chip to the implementers as the architecture is to the chip performance. Thus, changing the clocking strategy deep into the product cycle fundamentally sets the implementation back months, if not years.

Before embarking on a discussion of clocking styles, it is important to first lay some groundwork. This chapter first discusses the problems associated with clock systems, such as jitter, skew and noise generation. Clock generation and clock distribution, the building blocks of an on chip clock system, are then discussed. The clock distribution section includes a comparison of the industry's best clock distribution networks. Finally synchronous and asynchronous clock styles are discussed.

7.2 Clock Jitter and Skew

Clock generation and distribution are important considerations in high performance designs. Inaccuracy (uncertain movement) in the clock edge arrival times can have catastrophic effects on the logic and latching within the machine pipeline. The uncertainty may reduce the chip's performance or, worse yet, cause a functional failure [7.1].

The effects of clock uncertainty are well understood. For example, if the clock signal edge to an edge-triggered storage element[1] arrives earlier than expected, then the element may store the wrong information. Similarly, if the evaluation clock signal edge to a dynamic circuit receiving static signals arrives earlier and this occurs before the input signals have settled, then the circuit may improperly evaluate the input information. Furthermore, if a transparent latch[2] receives data from a dynamic circuit and the closing edge of the clock arrives later in time, then the latch may store the precharged data.[3] Consequently, managing and accurately predicting the inaccuracy of the clock edges is important for high speed designs.

Clock inaccuracy comes from two main factors: jitter and skew. Jitter refers to the inaccuracy that arises from the clock generation circuitry. Skew refers to the inaccuracy that arises from variations in the distribution system of buffers and wires to the individual circuits. The total clock inaccuracy is composed of these two components.

The study of clock jitter and skew is expansive and much too broad to present in detail here. As a result, this section is intended to give a brief overview of the problems associated with jitter and skew and their management.

7.2.1 Clock Jitter

Clock jitter is the inherent clock inaccuracy in the clock generation circuitry.

Clock jitter is the clock edge inaccuracy introduced by the clock signal generation circuitry. Consequently, there is little designers can do to reduce or prevent jitter, except for those who design the generation circuitry itself.

Jitter sources are dependent on the particular clock generation embodiment for the design. In today's designs, there are two main embodiments for clock generation, both of which center on maintaining synchronization with the system (board) clock.

---

1. For example, a master-slave flip-flop (see Chapter 5.2.1).
2. See Chapter 5.2.
3. Assuming the dynamic logic and latch receive the same clock phase.

The first type uses an on-chip phase-locked loop, more commonly referred to as a "PLL." The second type uses an off-chip oscillator with aligning circuitry inside the chip.

7.2.2 Clock Skew

Clock skew is the clock inaccuracy introduced by the distribution system.

Clock inaccuracy also arises from the distribution of the clock signal to the various chip components. Called clock skew, its amount is determined by several factors of the chip design. First, buffering the clock signal to obtain enough drive for the individual chip components introduces a component of inaccuracy due to process variations in both devices and wiring (see Chapter 1). An additional inaccuracy arises from the variations in the wiring of the various buffers across the entire chip design (see Chapter 1). A third inaccuracy arises from the potential unbalancing of the distribution system of buffers and wires. Additional inaccuracies arise from the environment surrounding the clock wires, that is, how much wire is adjacent to or above or below the wires or how the signals supplying these wires switch over time. Changes in the inductive reactance of the wires is also becoming significant with the increased use of wide clock wires. The reduced resistance of the wide wires and the high frequencies often used increase the inductive reactance. Differences in return paths as well as temperature and voltage variations across a chip can also cause uncertainties.

7.2.3 *Total* Clock Inaccuracy

Clock inaccuracy is the sum of jitter, determined by the clock source, and skew, determined by the distribution system, and can lead to a machine error or a reduction in the achievable chip frequency.

As mentioned previously the overall inaccuracy in the clock is composed of the sum of the jitter and skew components. The jitter term arises from the inherent inaccuracy in the clock signal generation while the skew arises from the buffering and distribution of the clock signal to the circuits. These two terms add together (rather than RMS) as they arise from two distinct sources and have little (if any) common source. The result is the inherent clock inaccuracy of the particular system.

> **Rule of thumb:** The clock skew and jitter form the inherent clock inaccuracy.

> **Rule of thumb:** The total clock inaccuracy (jitter & skew) is generally about 10% of the clock cycle time, but can range from about 4% to 13% depending on the clock generation circuitry and distribution system.

It is important to note that the effects of the clock inaccuracy are highly dependent on the design style of both the latches and circuitry as well as the clocking system. For example, if a design employs edge-triggered storage elements[4] and the clock signal arrives earlier than expected, then the given element will store the information present on the inputs at that time. If the input has not yet reached its proper state, then the wrong information will be stored, which creates a functional failure. Similarly, if the evaluation clock edge to a dynamic circuit arrives before the inputs have settled, a logical fault can also occur. That is, if the signal is in an illegal state when the circuit evaluation begins, then a false switching event will occur, creating a functional fail- ure. Note, though, that both of these problems can also be solved by providing more time for the input signals to settle, which can easily be achieved by slowing the clock signal. However, note that this means that the chip frequency is reduced.

Some negative effects of clock edge inaccuracy are not so readily counteracted. For example, consider the example of Figure 7.1 where a dynamic circuit feeds a trans- parent latch.[5] Note that both the logic and latch receive similar clock signals. If the clock signal falling edges are relatively close in arrival time, this means that the latch will always store the "Evaluate State." However, if the actual edge of the "Latch Clock" arrives after the "Logic Clock" has precharged the "Logic Output", the latch will store the "Precharged state." Slowing the clock signals will not fix this problem as both clock falling edges will be delayed, so the latch will always store the wrong information, which produces a machine error.

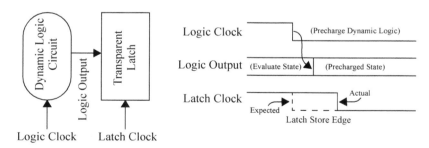

FIGURE 7.1 Latch Storage of Precharge Data Example

---

4. For example, a master-slave flip-flop (see Chapter 5.2.1).

5. See Chapter 5.2 (transparent latch ref)

The example of Figure 7.1 is really a complicated latch hold time violation.[6] In any machine, regardless of the circuits and latches, such a problem will always lead to machine error.

There are cases where the clock skew may not even matter (within reason). Chapter 7 illustrates the concept of slack borrowing where the logic style, latches, and clocking system all work together to produce a solution that is immune to a reasonable amount of clock skew. That is, in such a design the clock skew does not reduce the machine frequency.

Before turning our attention to actual clocking styles, the concept of noise generation by the clock distribution should first be discussed. That is, noise should be an additional consideration when selecting the clock distribution.

7.2.4 Clocks and Noise Generation

If not designed properly clock buffers may produce rail span collapse or (at the very least) variance; signals switch every cycle at sharp edge rates, increasing the possibility and severity of capacitive coupling.

Clock buffers and signal wires are bad neighbors: clock buffers tend to be large, drawing heavy current; clock signals switch every cycle, both up and down, causing potential coupling problems. Thus, both tend to proliferate the overall chip noise problem. Consequently, it is important to consider their effects when selecting both a clocking style and a clock distribution system.

Because they must ultimately produce signal(s) with large drive capability, clock buffers tend to be large, at least somewhere along the buffering tree. Consequently, they draw heavy current, necessitating a large power supply bus system, complicated power distribution analysis or a relaxed design to meet the power distribution requirement [7.14] . However, even the best power bussing system may not be sufficient to prevent a significant rail span collapse in the immediate area of the buffer due to inductive effects. If a noise coupling event to a sensitive circuit in the immediate area receives an input at such a time, then the induced supply/ground offset may be sufficient to cause a noise failure.[7] Even if a noise failure does not occur, a large current surge will at a minimum cause the power supply to ring. Also due to inductive effects, this ringing will eventually dampen out, but will reduce the accuracy of the circuit

---

6. See Chapter 5.2.3 for a detailed explanation of latch hold times.

7. For example, in the case of a domino circuit, a ground offset on the order of a threshold will cause a low value to be interpreted as high.

Clocking Styles   **251**

delays: the power supply can be at most guaranteed to be the lowest possible value due to the ringing, which reduces the circuit performance.

Additionally, the output of the clock buffers are unfriendly as well. Each signal switches both up and down every cycle at (generally) a low transition rate, which increase the amount of coupling to adjacent nets.[8] Because clock wires tend to be long, the probability of a wire being routed adjacent to a given clock net is high.[9] Furthermore, the more clock nets routed, the higher the probability of a clock net coupling into a net of questionable noise margin.[10] Thus, the clock distribution system design must also account for the possibility of such problematic events and protect the various circuits accordingly. A solution is to shield the major trunks of the clock nets on one or both sides with power supply lines to prevent such problem coupling events.[11]

When considered, the noise ramifications may alter one's view of the ideal system. That is, a clock distribution system that is good for clock skew, such as the water-main approach, may actually be worse in the long run due to supply variation issues. Or, likewise, a grid system may create such an additional wireability impact (if shielded) or a large number of hostile neighbors (for noise) that the advantages of the system (e.g., predictability, skew) are quickly negated.

7.3 Clock Generation

7.3.1 PLL Based Designs

Synchronization within a computer system is typically accomplished using Phase-Locked Loops (PLL). A low speed reference clock is distributed throughout the system to each chip to be synchronized. A PLL on each chip multiplies the reference clock by some fixed value, i.e. 1.5, 2, 2.5, etc. The reference clock serves as both a frequency reference and a time zero phase reference for the system. Typically the internal chip clock operates at a multiple of the reference clock and the IO circuits operate at or near the frequency of the reference clock.

---

8. This is because of dV/dt effects.

9. That is, a long wire has a higher probability of a net occupying one of the adjacent (horizontal) tracks.

10. For example, a domino input.

11. Note, too, that this has good effects for inductive return paths.

The problems associated with a PLL based system are numerous. PLLs commonly employ analog circuits which are sensitive to noise and difficult to model. As power supply voltages decrease with more advanced technologies, designing noise insensitive PLLs is more difficult. PLLs require AC device models in order to be able to accurately predict PLL characteristics and performance. The circuitry inside the PLL has an inaccuracy because of the statistical nature of the affecting process parameters. Additionally, the PLL circuit uses a lock-in technique to determine the precise clock frequency of the incoming board-level clock signal.However, the granularity of the lock-in circuitry is finite, only able to adjust in discrete amounts.Thus, the clock signals are only accurate to the degree that the PLL circuit is able to discern. Clock jitter is this inaccuracy added to that arising from process variations for the PLL circuitry

> **Rule of thumb:** Clock jitter in a PLL based design is roughly 5% of the clock cycle time.

PLL based designs are extremely useful for systems requiring chip frequencies that vary widely with the system bus. Examples of designs using this technique are the IBM PowerPC™ and Intel Pentium™ microprocessors.

A PLL is used to multiply an external reference clock to produce or "synthesize" the on chip clock.

The PLL multiplication factor is often programmable to allow the chip to operate over a wide range of internal cycle times.

The PLL drives the main clock distribution on chip to the majority of the latches.

The PLL provides latency correction for the delay through the chip's clock distribution

Any jitter (variation in cycle to cycle clock period) on the PLL will impact the maximum speed of a chip.

The main chip clock is often divided down to generate the IO or Bus clock.

The bus clock is used to latch the input data or drive output data to other chips in the system.(The PLL phase aligns the Bus or IO clock with system reference.)

FIGURE 7.2 Summary of PLL characteristics.

7.3.2 Off-Chip Oscillator Based Design

Off-chip oscillators are alternatives to a PLL based design. Typically, high frequency oscillators are very stable and have low jitter. Designs using an off-chip oscillator, however, rely on board-level synchronization. That is, there is no on-chip circuitry (such as a PLL) to insure that the internal clock stays synchronized with the system clock. Consequently, the only source of jitter in such a design arises from the accuracy of the off-chip oscillator, which is a minimum of only 1% in some cases, but rarely (if ever) as much as 5% of the clock cycle time.

> **Rule of thumb:** Clock jitter in a off-chip oscillator based design is typically less than 5% of the clock cycle time, and can be as low as 1%.

Off-chip oscillator designs, however, push the synchronization of the component chips to the system board, which can be more expensive and difficult. Additionally, systems that cannot run both the chip and system bus at the same frequency have difficulty using this technique, particularly if the chip to bus clock ratio is not an integer. An example of a design using this technique is the Digital Equipment Corporation (DEC) Alpha™ family of microprocessors. [7.2]

7.3.3 Delay Locked Loops

A Digital Delay Locked Loop (DLL) can be used to correct for the latency of the clock distribution or to set the phase of the bus clock relative to the internal chip clock. In general, DLLs are easier to design and analyze compared to a PLL. A digital DLL can be designed using almost entirely standard Engineering Design Systems (EDS) tools and circuits, i.e. state machines.

> **Rule of thumb:** When should one use DLL vs. PLL? As mentioned earlier, a PLL requires analog design, modeling and skills. PLLs tend to be more sensitive to noise, however, when properly designed offer better phase resolution. PLLs are almost always employed for frequency synthesis. DLLs are most often used to perform simple phase adjustments of the clock, or for latency correction.

7.3.4 Phase Splitters

Clock phase splitters or true complement generators take a single clock phase as an input and generate two separate phases as outputs. One output phase is usually a buffered version of the input and the other phase is an inverted (180 degrees out of phase) and buffered version of the input. Splitters are a key building block for two phase design styles. The key metrics are size, transition time, and gating features. Size is important if the splitter must fit into a bitstack or data flow pitch, plus the number of splitters on a chip can be large. Splitters should provide fast transitions to the latches since this transition affects the overall performance. It is often desirable or necessary to gate splitters for testability reasons as well as for power savings or functional gating. Placing a splitter at the beginning of a tree would require two distributions after the splitter; this will greatly increase skew. On most high performance designs, splitters are placed at the end of the tree to save wiring and reduce skew by keeping both phases common for as much of the distribution as possible.Splitters can be designed with different amounts of overlap or programmable overlaps which can be used to gain or reduce margin on well-characterized clock paths; this will help performance. Phase splitters should be tuned to eliminate mid-cycle clock skew (skew from master to slave latches) caused by on-chip process variations [7.16] .

> **Rule of thumb:** Splitters should be placed at the end of a distribution, not in the beginning.

7.3.5 Clock Chopping

Clock chopping is the process of taking a single edge as an input and generating multiple edges (pulses) from that signal. The circuit techniques for doing this are called schmitt triggers of monostable-multivibrators or clock choppers. Clock choppers are used for frequency multiplication or when a library element requires a specific timing or pulse width (e.g., RAMs). A concern with clock choppers is how to test these with a static scan based test (i.e. how to observe the input and control the output). Because of this issue, usage of clock choppers is often limited to well characterizable parts of the design.

7.4 Clock Distribution

Buffering the main clock signal(s) typically involves building up signal gain from the base clock source[12] to drive a large central clock buffer. This buffer typically resides in a central location to drive a series of strategically located buffers for particular chip areas. Several more stages of buffering may be involved before the clock signal finally arrives at individual circuits. Figure 7.3 shows a sample clock buffering tree schematic. Note that the tree uses n stages of buffering and produces m distinct clock outputs.

Due to statistical process variations, each buffer stage in the clock system produces a component of inaccuracy (due to the device variations of the process) to the clock arrival times relative to each parallel stage in the system. This inaccuracy adds as buffering stages are increased to obtain the necessary drive to distribute the clock across the chip. Consequently, the depth of the buffering must be kept to a minimum

---

12. For example, a PLL.

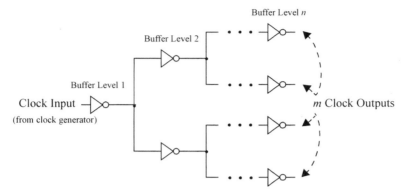

FIGURE 7.3 Sample Clock Buffer Tree Schematic

to reduce the amount of clock variation (skew) introduced by variations between clock buffer stages.

> **Rule of thumb:** The number of buffering stages in the clock system should be minimized to reduce the skew created by process variations.

As is apparent in Figure 7.3, the clock buffering stages are interconnected using chip wiring. As Chapter 1 described, the wiring levels are governed by statistical process distributions (like the FET devices) in length, width, thickness, and sheet resistance as well as dielectric thicknesses and permittivities. These variances mean the capacitance and resistance of the interconnecting wires follows a statistical variation, which has little or no correlation to the device distributions.[13] Longer wire runs introduce more RC effects into the delay between buffer stages, which can compound the level of uncertainty: resistance uncertainty must be combined with that of the capacitance. Additionally, wiring capacitance is also affected by the switching characteristics of adjacent neighbors, which are more pronounced for longer wires.[14] Consequently, distributing the buffer stages throughout the chip to reduce the interconnect lengths is advantageous: shorter runs reduce RC delay effects. Figure 7.4 on page 257 shows a sample clock buffer placement diagram.

> **Rule of thumb:** Clock buffer stages should be scattered across the chip to avoid large RC effects by reducing interconnect lengths. This of course gets tricky since the pre-

---

13. As noted in Chapter 1, the device (FEOL) and interconnect (BEOL) process distributions do not track with one another.

14. Wire capacitance can be affected by as much as 2x. Since longer wires means more delay in the wire, this directly translates into more wiring uncertainty.

vious rule wants to reduce the number of stages.

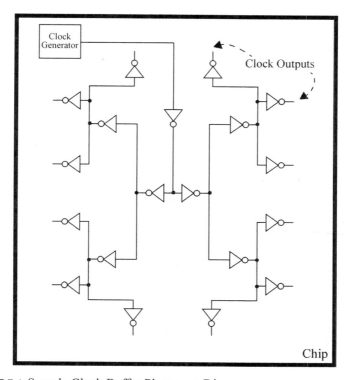

FIGURE 7.4 Sample Clock Buffer Placement Diagram

A third effect that adds to clock skew is the inevitable unbalancing of the buffering system. That is, each circuit that requires a clock, whether a dynamic logic block or a storage element, creates a load on the clock system. These are grouped and driven by a buffering stage which must be sized according to the particular load. Each such buffering stage produces, then, a different load to its driving signal. This difference cascades back through the clock buffering hierarchy, creating a load balancing problem at each stage of the system. This problem must be solved with buffer placement along with wires to force the clock signals at each buffer level to arrive as close as possible in both delay and transition time, else clock skew is introduced. Realistically, a zero delta in either time or transition across all buffers in all stages is unachievable. For example, on a superscalar microprocessor of 15+ million transistors using tens of thousands of registers and dynamic circuit blocks and 5+ levels of clock buffering stages, it is absurd to expect a balanced clock buffering system.

> **Rule of thumb:** An unbalanced clock buffering system is inevitable.For the average distribution system in a high speed design, clock skew ranges typically from 3% to 8% of the clock cycle time, with the average design at about 5%.

7.4.1 Clock Distribution Techniques

There are many clock distribution techniques, each with advantages and disadvantages, that should be considered for a high speed design. Most importantly, the clock distribution system should be finalized up-front in the design cycle and not revisited.

Some first order metrics used to compare various types of distributions are: skew, wireability, rail collapse, integration ease, and automatic routing.

There are six basic clock distribution techniques in use in today's VLSI chip designs. The first is distributing buffers using some placement optimization and standard wiring but using distributed buffers to control the delay. The second is the "water-main" technique where a central clock buffer provides nearly all the buffering. The third involves creating a clock "H-tree" where the loads at each buffer and wiring stage are finely tuned/balanced. The fourth type is the "grid system" where the clock is driven throughout the chip across a predetermined grid of wires. The fifth uses serpentine wires or added wire length to balance delay. The last type is a hybrid approach which tries to meld the best characteristics of all other approaches and minimize their faults.

7.4.2 Distributed buffers, placement optimization and standard wiring

The first method is most flexible and easiest to automate and is common in many ASIC chip designs because of its flexibility. It can handle many clock domains which is critical to many ASIC chips [7.14] [7.15] . Its advantages are flexibility, wireability and low power usage. Although design skew can be contained by sophisticated placement optimization and balanced routing, process skew is very high because of the many buffer stages.

7.4.3 Water-main Clock Distribution Technique

The water-main approach to clock distribution is very simple. In this technique, the clock signal from the generation circuitry is buffered to gain a large amount of drive for the entire chip as shown in Figure 7.5. Note that this system is identical to that of Figure 7.3 on page 256 with the exception that a single clock output is required rather than several.

FIGURE 7.5 Water-Main Clock Buffering Schematic

The main clock signal is routed across the chip via a very wide wire in the center of the chip, typically either in a strip or a cross configuration.[15] Figure 7.6 shows an example strip configuration.

Note that the selection of a strip or a cross design is dependent on the shape of the overall chip. That is, if the chip is shaped like a narrow rectangle, then a strip is perfectly suitable. Likewise, if the chip is more square in nature, then a cross design is more advantageous to provide an adequate clock signal to the circuitry.

Thus, the name "water-main" arises from the fact that a large "pipe" feeds the clock to the entire chip. Individual circuits that require the clock signal do so by using "taps" to the main line. Thus, the entire system is analogous to the distribution of water across a large system.

When used, the clock skew in a water-main system follows a contour across the chip based on the distance from the feed. That is, the system creates a set of contour lines around the main clock line, each representing a particular delay (skew) from the main clock based on locality. Figure 7.6 on page 260 shows the clock skew contour lines for a sample strip configuration.

It is possible to take advantage of low horizontal skew in this case by using horizontal data flow for logic function. This would allow better skew within the horizontal data flow units.

7.4.4 H-Tree Clock Distribution Technique

The term "H-tree" arises from the nature of the clock buffer placement and wiring. Note in Figure 7.7 that the wires between buffer stages are configured in a balanced "H" wiring pattern from the source to the sink points. Buffers can be placed at the end of the "H" pattern.

---

15. There are other possibilities for routing the main clock, such as a star configuration, etc. However, most rely on the strip or cross configuration and few (if any) designs do otherwise.

Clocking Styles **259**

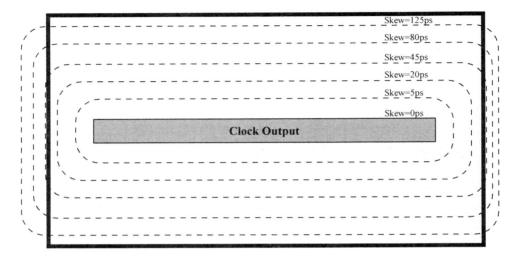

FIGURE 7.6 Water-Main Clock Routing: Strip Configuration Skew Contours

Balancing the H segments in an H-tree system can be quite difficult. That is, each wire in the system must be balanced to produce the same RC or RLC delay, else an immediate component of clock skew will arise. This means that the wires must be extensively tuned based on the load at each buffering stage. Consequently, because each circuit receiving a clock signal presents some amount of load at the last clock buffering stage, the drivers at each point in the clock tree are not uniform. The result is that the clock H-tree balancing problem cannot be performed until the clock loads are already known and the chip-level integration has a finished placement. However, once completed, the balanced H-tree clock system has a uniform clock skew based on the specified target used in the design of the buffering system,[16] remembering that the process tolerances will limit the accuracy as well.[17] To aid in the design of such trees, tools are available to automatically create the H-tree segments.

However, in a buffered H-tree system, the various branches yield a unique situation to the clock skew problem. The clock skew between any two logically connected circuits at the final buffering stage varies depending on how common their distributions

---

16. The target of the clock skew must be as small as reasonably possible, but must also be realistic in terms of schedule and cost. That is, finer accuracy requires more time for completion.

17. For example, the ACLV tolerances for both front-end and back-end process components (see Chapter 1).

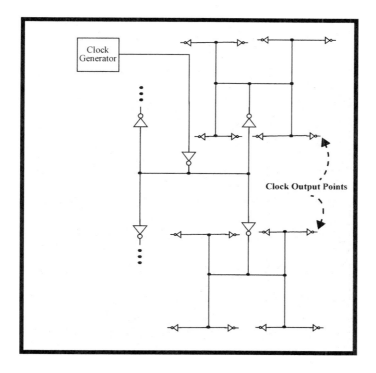

FIGURE 7.7 H-Tree Clock Routing Diagram

are. That is, if they both receive their clock signals from the same final clock buffer, then the skew between them is lower than if they receive them from separate buffers at the next-to-last stage. Figure 7.8 illustrates this point for a sample three-level H-tree buffering system.

	Same Third Level	Same Second Level	Same First Level
Source-to-Sink Clock Skew	25 ps	150 ps	260 ps

FIGURE 7.8 Sample H-Tree Clock Distribution Skew Chart for a 3-Level System

> **Rule of thumb:** H-tree clock routing is excellent for chip wireability and rail collapse, but poor for automatic clock routing.

7.4.5 Grid Clock Distribution Technique

The grid approach to clock distribution tries to solve the routing problem caused by the water-main technique by creating more points for connection like the H-tree system. Another goal is to allow the clock system wiring to be established at the design outset, rather than waiting for the circuit placement information to be completed as in the H-tree technique.

The grid technique uses a clock buffering schematic system similar to that in Figure 7.3 on page 256, as in the H-tree system, and a buffer placement strategy as in Figure 7.4 on page 257. However, this scheme *may* tie the output of one or more intermediate buffering levels together to reduce the clock skew.[18] At the last (n^{th}) level of buffering, the outputs of the clock buffers are also tied together and routed throughout the chip in perpendicular wiring levels at a set periodicity, connected at each crossing point. See Figure 1.12 for an example of a grid clock distribution.

Clock skew control in a grid system is somewhat similar to that of the H-tree technique in that the skew goal of the entire system can be set up-front. However, it improves the overall skew because the buffering is more like the water-main technique in having a central source. Note, however, that this central source is actually made up of distributed buffers across the chip. This means that although the outputs of the final buffer stages may be tied together via the clock grid, the device and wiring (RC) process variations will increase the overall skew beyond that of the water-main system.

Section 1.4.2 discusses a grid clock distribution in greater detail.

> **Rule of thumb:** Grid clock routing is excellent for automatic routing, but poor for chip wireability.

7.4.6 Length-matched Serpentines.

Length-matched Serpentines is a technique employed by the Intel P6 [7.8] . It is described in section 7.4.8 .

---

18. Tying the buffer outputs at each level reduces skew by allowing stronger buffers to help weaker ones as introduced by processing tolerances (see Chapter 1). However, it also introduces additional wiring complications due to the routing of additional clock lines.

7.4.7 Hybrid Clock Distribution Techniques

It should be somewhat obvious from the previous discussions that no single solution to clock distribution is ideal in itself. Consequently, many designs seek to marry the positive attributes of the various techniques to create a substantially better distribution solution than any single type. Many of these techniques are applied at each of the clock buffering levels, resulting in a multitude of hybrid approaches.

> **Rule of thumb:** Hybrid clock routing tries to marry positive aspects of the various approaches, while minimizing the negative aspects.

7.4.8 Comparison of the Industry's Best Clock Distribution Networks

Clock distribution has become an increasingly challenging problem for VLSI designs, consuming an increasing fraction of resources such as wiring, power, and design time. Unwanted differences or uncertainties in clock network delays degrade performance or cause functional errors. Three dramatically different strategies being used in the VLSI industry to address these challenges are compared. Novel modeling and measurement techniques are used to investigate on-chip transmission-line effects that are important for high performance clock distribution networks.

The rapid increase in clock frequency and transistor count poses many problems for VLSI designers. One challenge receiving increased attention is the seemingly simple task of supplying one or more clock signals to all the latches and clocked dynamic gates on a chip.

Interconnects have become more important due to fundamental scaling effects that increase the fraction of the chip cycle time that is consumed by interconnect parasitic capacitances, delays, and coupling effects. Significant effort is devoted to modeling and reducing these delays through design optimization and technology improvements. Using these design and technology advances with careful micro-architecture and chip organization, it is still possible for experienced design teams to keep interconnect delays a small fraction of critical path delays. Clock distribution, however, is unique in that the total delay of the clock network is already dominated by interconnects. Buffer delays and capacitance have scaled to keep up with cycle time, but interconnect delay from the center of the chip to corners has not, while the number of clocked gates continues to grow. Thus clock distribution problems lead to unique modeling and design techniques as well as technology advances.

While the delay of a clock distribution network is relatively unimportant, any modeling error or uncertainty in the clock signal arrival times between key points in the clock distribution can cause performance or functional problems.

To reduce model, process, and noise induced clock distribution uncertainties, the total delay through the clock distribution is, in general, minimized. This leads to the use of long, wide wires placed on the lowest resistance wiring levels, that are driven with fast transition times, which in turn leads to significant transmission-line effects [7.4] . Uncontrolled transmission-line effects are a growing source of uncertainty and clock skew as will be discussed below.

7.4.9 Network Topologies

Most high performance microprocessors distribute a single performance-critical clock signal to many locations on the chip, although local regions may be gated for power management. The different local clock phases needed for various latch circuits, arrays, or dynamic logic are then generated locally from this global clock signal. There is a wide variety of clock distribution network topologies now being used for global clock distribution.

Simplified electrical models very roughly inspired by three commercial microprocessor clock network topologies were studied for illustrative purposes: *grids* like the DEC 21264 [7.5] , *trees* like the IBM S/390 [7.6] [7.7] , and length matched *serpentines* like the Intel P6 [7.8] . The goal of the comparison is to understand the advantages and disadvantages of these very different topologies, (without reproducing many important details) by creating simple simulation models of each topology. The effects of wiring technology will be simulated, considering Al and Cu wires, with and without on-chip dedicated reference layers. The implications of non-ideal real-world cases will be discussed, followed by design and measurements of a 400MHz product.

All three microprocessors use tree-like networks driving roughly 16 buffer or spine locations for the first, longest wires in the global clock distribution. This consensus occurs because perfectly symmetric H-trees driving identical loads result in zero nominal skew. The very different networks driven by these buffers will be the focus of this section.

Each of the simplified topologies is designed to drive only one quadrant of a 17 x 17 mm chip containing 150 pF of uniformly distributed load in each quadrant. In figures showing physical wiring, all wire widths are drawn 10X wider to allow better visual comparison of wire widths.

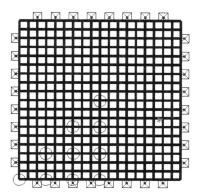

FIGURE 7.9 Clock grid for a chip quadrant, crossed boxes show drivers, circles show locations of simulated waveforms.

7.4.10 Grids

The DEC Alpha series of chips uses grid-based clock distributions driven by one or more lines of buffers. This robust topology guarantees very low skew in any local region, and can be routed early in the design. Figure 7.9 shows a grid based network for one quadrant of the DEC 21264, driving 150 pF of gate load, using 350 pF of grid wire capacitance. The number of grid wires was chosen arbitrarily, then a wire width of 14 μm approximately reproduced the published wire capacitance [7.5] . The DEC process includes reference planes above and below the two planes used for the clock grid wiring [7.5] . Figure 7.10 shows simulated waveforms for this grid

Figure 7.11 shows the expected increase in transmission-line effects if the Al reference-layers were instead used as standard orthogonal wiring levels, and the V_{DD}/Gnd return conductors were instead routed on both sides of each grid wire. Due to the increase in the average distance to the return current path when the reference planes are removed, the inductance increases, and the transmission-line effects such as plateaus and reflections become more pronounced, increasing the clock skew. These transmission-line effects can be reduced by using a finer grid having a larger number of narrower wires, but this increases total capacitance, wire delay, and skew.

A modified grid was also simulated assuming a Cu wiring technology [7.9] , that does not use dedicated reference planes (Figure 7.12).

For the Cu technology, twice as many 4.5 μm wide grid wires were used, with reduced driver sizes. The thinner, narrower, Cu wires result in reduced power (C_{wire}

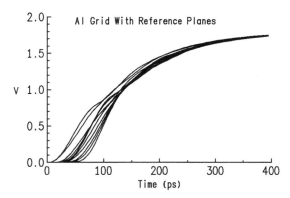

FIGURE 7.10 Simulations of clock grid of Figure 7.9 using Al technology with reference planes.

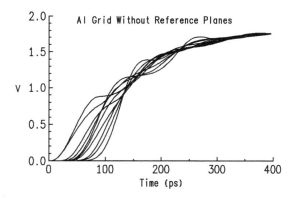

FIGURE 7.11 Simulations of grid of Figure 7.9, but without reference planes for low inductance paths.

reduced from 350 pF to 228 pF) even with the finer grid, and exhibited no need for dedicated reference planes.

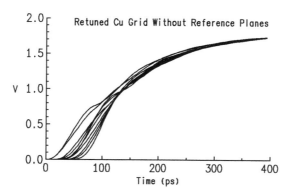

FIGURE 7.12 Simulations of clock grid redesigned for copper grid with twice as many narrower wires and reduced driver sizes.

7.4.11 Trees

Figure 7.13 shows a symmetric H-tree designed to drive the same 150 pF loads as the grid in the previous section, with the maximum wire width chosen to be (the same width used for the Al grid) but (as shown) wire widths were optimized for minimum delay. Due to assumed idealized symmetry of the loads, the simulated skew is virtually zero.

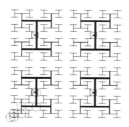

FIGURE 7.13 Tree topology driving same loads as grid above.

Figure 7.14 shows significant transmission-line effects at various internal nodes within the trees, but only the smooth signals at the ends of the trees are relevant. Any overshoot can easily be controlled by reducing wire widths, and may be desirable as it is accompanied by faster transition times.

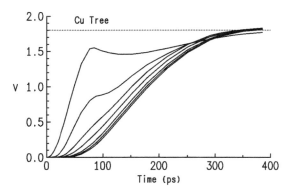

FIGURE 7.14 Simulations of copper tree in FIGURE 7.13 .

7.4.12 Length-Matched Serpentines

Figure 7.15 shows another topology where each load is driven by a single point-to-point wire, and lengths are matched using a serpentine structure. To achieve delays and transition times similar to the trees, wire widths were chosen to be 1.6 μm.

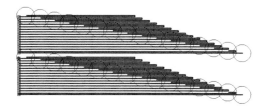

FIGURE 7.15 Serpentine wires driven by a line of clock drivers at the left edge. Only 32 of the 256 serpentines needed to drive the 256 loads of grid and tree examples are shown.

This topology is relatively simple to design, and like symmetric trees, has virtually zero skew for identical loads, as long as coupling and uncontrolled variables are insignificant.

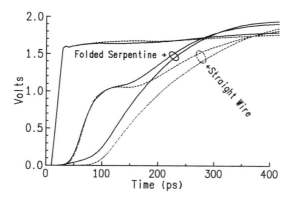

FIGURE 7.16 Simulated wave-forms at the start, middle, and end of straight vs. folded identical length wires of 1cm length and 5.3 μm width driving 2pF. Grounded 1.2 μm shield wires provide capacitance shielding but inadequate inductive shielding (with no reference planes.)

7.4.13 Comparisons

Table 7-1 compares the three topologies assuming uniform load distribution and no environmental or process variations for the Cu wiring simulations (with no reference planes).

	C_{wire}	Delay	Skew
Grid	228 pF	21 ps	21 ps
Trees	15.5 pF	130 ps	0 ps
Serpentines	480 pF	130 ps	0 ps

TABLE 7-1 Table 1: Uniform Load Distribution

For these symmetric cases, the tree topology provides low skew and much lower capacitance than the other topologies, although it requires placement of buffers at four locations internal to each chip quadrant. For real designs, a number of complications arise that further differentiate the topologies. First, actual capacitance of individual clock pins can vary from a few fF to a few pF at each pin. In addition, the loads are distributed non-uniformly over the chip, and across-chip process and power supply variations can be significant. One major advantage of the grid topology is that even very non-uniform load distributions affect the local skew very little. Thus, changes in clock loads, locations, or electrical models cause little change in clock timing, and rarely require re-tuning of the grid wires or drivers.

Although trees are potentially more efficient, wiring and tuning tree topologies to drive highly non-uniform loads with low skew can be much more difficult. Modeling errors or process variations can produce large skew even between nearby clock pins. An example is discussed below.

The serpentine structure is simple to design for different load locations; however, the number of serpentine wires is large and changeable, contributing to wire congestion. Small loads require capacitance padding, while wide serpentine wires used to drive larger loads can lead to inductive self-coupling (Figure 7.16). Since a full-chip distribution using this method requires a line of drivers on the left and right chip edges, there is potentially large skew between the left and right halves of the chip due to cross-chip variations [7.8] .

7.4.14 A 400 MHz Clock Tree Design

To take advantage of the efficiency, flexibility, low power, and potentially low skew of the tree structure, a proprietary low skew clock routing tool was developed to drive arbitrary load distributions with arbitrary wire widths while avoiding blockages. An optimization process also considers power, wiring tracks, and process variations. The tool has been used on several IBM microprocessor and ASIC designs [7.14] . Measurements were made using an e-beam system [7.7] , and backside photo-emission [7.10] (not shown).

Figure 7.17 shows the importance of including transmission- line effects in the design of high-performance clock distributions. Although the topologies and loads were similar, a product chip designed with transmission-line effects included in the routing, extraction and tuning algorithms showed a 5X reduction in skew compared to the test chip designed without these considerations [7.7] .

Figure 7.18 shows a unique representation of the product's clock-tree network. The clock routing tool matches delays, but for efficiency does not match lengths or loads, so subtle differences remain.

Figure 7.19 shows that clock pins at the ends of the longer trees exhibited more measured and simulated overshoot due to the faster signal speed. The magnitude of the overshoot is adequately modeled by the frequency independent inductance and resistance model used in the design-tools (Figure 7.19), but a frequency-dependent model

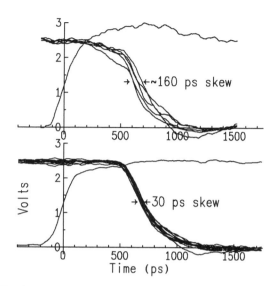

FIGURE 7.17 Hardware e-beam measurements. The top shows a test chip designed with RC wire models, the bottom shows waveforms from the 400Mhz microprocessor clock network designed with transmission-line considerations. The rising signals are the output of the single central chip buffer, while the falling waveforms are 10 of 580 clock pins.

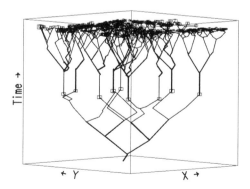

FIGURE 7.18 A 3-D **(X, Y, Time)** representation of the microprocessor tree network distributing the global clock to 580 local clock pins. The signal starts at the bottom, and the 9 vertical lines represent buffers. Gate loads are represented by cubes with volume proportional to capacitance

(not shown) including the extended wiring environment is needed to match the details of the measurement.

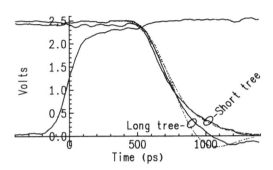

FIGURE 7.19 The measured (solid) and simulated (dashed) waveforms from the shortest and longest wiring trees shown in figs 9 and 10.

7.4.15 Clock Distribution Techniques Summary

In summary, each clock distribution technique has its various advantages and disadvantages. Because of these, the selection of the style of distribution system depends heavily on which aspect(s) the chip design team has deemed most important. For example, if clock skew outweighs chip wireability, then the water-main or grid system is most appropriate. Alternatively, if wireability is the more important, then an H-tree system may be more advantageous. Or, potentially, if all factors are equally important, a hybrid system may be the best solution.

In any case, the distribution technique should be decided and designed up-front, before the actual design is implemented. This permits the implementers to design their circuitry with the particular clock skew tolerances in mind, thereby counteracting the effects as much as possible. Therefore, because the final clock skew is rarely known down to the last picosecond at the outset, an accurate estimate must be provided.

Furthermore, in any distribution style, it is important to keep the clock edges as crisp and sharp as possible. This reduces the effects of wide variances on the FEOL/BEOL process sectors ("Back-End-Of-Line Variability Considerations" on page 36), which directly translates into lower clock skew, regardless of the clock distribution system.

7.5 Single Phase Clocking

In single phase clocking, all circuits receive an identical clock signal.

In single phase clocking, a single clock signal is distributed throughout the chip using a buffering system like that of Figure 7.3 or Figure 7.5. This results in the circuits receiving a clock signal identical to that from the clock generator, except for a time delay incurred by the buffers and wires along the distribution network, as shown in Figure 7.20.

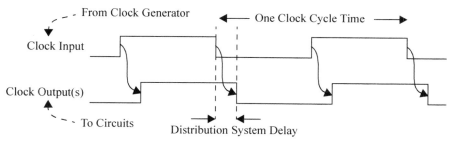

FIGURE 7.20 Single Phase Clock Diagram

Single phase designs use one of three basic types. The first is a master-slave approach where master-slave flip-flops trigger are triggered by the same clock edge, separating them by a full clock cycle. The second strategy uses latch elements triggered off opposite edges of the clock, which separates each latch element from the next by one-half of a cycle. Lastly, some systems use alternating nmos/pmos dynamic logic, which results in data storage at each circuit.

7.5.1 Single-Phase Master-Slave Design

In single phase master-slave designs, logically adjacent master-slave flip-flops are separated by full clock cycles, which necessitates the use of static logic.

Single phase master-slave design is the easiest style of all to both understand and implement. It uses only single-clock master-slave flip-flops and static logic connected as shown in Figure 7.21. This produces a simple, but robust design.

As described in Chapter 5.2, each flip-flop in the system stores the information on its input when the storage edge of the clock arrives. Thus, the time between any two log-

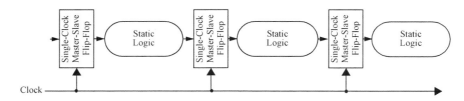

FIGURE 7.21 Sample Master-Slave Design

ically adjacent flip-flops is equal to one clock cycle time. This is important to note because the logic between elements cannot hide a precharge cycle. That is, if a pre-charge cycle is required, then this directly reduces the active from the cycle time. For example, consider the case where a domino circuit is placed between two positive edge-triggered flip-flops. The output of the first flip-flop arrives on the rising edge of the clock, which is the input to the domino. Assume the domino circuit uses the clock in an inverted form and evaluates while the clock is low and precharges while the clock is high. Thus, it will be in precharge when the first flip-flop output is switching, preventing the problem of an incorrect evaluation. When the clock is low, the domino evaluates the input data and the result appear on the output. On the next rising edge of the clock, the second flip-flop will (properly) store the evaluated state. Note that the domino circuit will now precharge again. Referring back to Figure 7.20, this means that the time while the clock signal is high cannot be used for logical evaluation and is lost in the precharge. Thus, the precharge time for dynamic logic cannot be hidden using such a system and directly reduces the performance of the machine.

If the system is changed such that the domino evaluates on the rising edge of the clock, then its input will not be settled as the flip-flop output is not settled. Thus, a premature evaluation will occur unless the clock to the domino is delayed beyond the time the output has stabilized. Assuming this problem has been solved, note that the domino circuit will precharge its output, which is the input to the second flip-flop, on the next falling edge of the clock. Thus, when the next rising clock edge occurs, the second flip-flop will incorrectly store the precharge state.

Therefore, it is important to note a single-phase master-slave design system[19] necessi-tates the use of static logic. That is, precharge time must take a portion of the cycle

---

19. And, also, a two-phase master-slave system as well - see Section 7.6.1 on page 281.

time, which reduces the time the machine has to perform real operations. This leads to a performance penalty.

> **Rule of thumb:** Single-phase master-slave design cannot bury a precharge cycle and, thus, necessitates the use of static (non-clocked) logic to prevent an associated performance penalty.

Flush-through in such a system is only a remote possibility. That is, the master-slave flip-flop can lose the machine state if both the master and slave portions receive data simultaneously. However, because only single-clock master-slave flip-flops (see Chapter 5.2.1) are used, such problems are isolated to the flip-flops and are not exposed to the clocking system.

Clock skew tolerance in a single-phase master-slave design is poor. Each storage element is edge-triggered, which means that if the clock to a given flip-flop arrives early due to skew, the wrong information will be stored. If the clock arrives late due to skew, the proper information will be stored, but the overall chip performance will be degraded by the skew amount. That is, the cycle time of the machine will be

$$t_{cycle-actual} = t_{cycle-desired} + t_{skew}. \qquad (2.1)$$

Using the fact that frequency is the inverse of the cycle time, this means that the actual achievable chip frequency is degraded to

$$f_{cycle-actual} = 1/(t_{cycle-desired} + t_{skew}) < f_{cycle-desired}. \qquad (2.2)$$

The net effect is that the clock skew must be accounted for in the design and added to the cycle time target, reducing the resulting frequency.

Furthermore, each single-phase master-slave flip-flop incurs a large delay penalty. From Chapter 5.2.3, each flip-flop requires a set-up time to the clock signal as well as a launch delay to produce the output. Recall that this time is relatively large compared to a flushing latch circuit (see Chapter 5.2). Both of these reduce the remaining time permitted for logic evaluation. That is, the amount of time available for the logic circuitry is the cycle time less the flip-flop set-up and launch time, or

$$\text{Available Time for Logical Evaluation} = t_{cycle} - t_{set-up} - t_{launch}. \qquad (2.3)$$

The net result of the clock skew and latch delay penalty problems is that single-phase master-slave designs are not suited for a high speed applications.

However, in its defense, the single-phase master-slave design style has a key advantage of design simplicity. That is, because there is only one logic type of interest and one storage element type, the number of possible combinations of logic and flip-flops is only one. From Figure 7.21, logic must feed (finally) a flip-flop and a flip-flop can drive only logic or another flip-flop. The result is that a single-phase master-slave flip-flop system is easy to design and analyze.[20]

In summary, a single phase master-slave design is *not* a viable high speed solution and is (generally) not used in critical circuit areas.[21]

> **Rule of thumb:** Think twice before using the single phase master-slave style.

7.5.2 Single Phase Separated Latch Design

<u>In single phase separated latch designs, logically adjacent latches are separated by half-cycles. Depending on the connectivity of the clock to the logic and latches, it permits the use of dynamic or both dynamic and static logic.</u>

Single phase separated latch design refers to designs where opposite (positive and negative) clock level activated d-type storage elements are cascaded with intervening logic. A base diagram is shown in Figure 7.22.

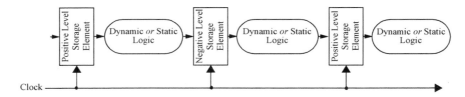

FIGURE 7.22 Single Phase Separated Latch Design Base Diagram

Referring back to Chapter 5, note that a master-slave flip-flop essentially consists of two adjacent d-type latches. Thus, the system of Figure 7.22 may be viewed as a master-slave design where the individual latch components are separated by intervening logic. This is the source of the term "separated latch design." Some readers may be also acquainted with the term "single latch design," which (typically) refers to the case where latches are not required to have intervening logic. Although very similar, the difference between these two is subtle (but also quite important): if the clock signals overlap significantly, due to either clock skew or by design then the single latch style will result in a flush-through problem. It is important to realize that the single

20. Actually, this is true of any style using only master-slave flip-flops.

21. It is possible to use such a style in non-critical circuit areas/paths within a style that is more conducive to high speed design as long as the clocking style is consistent. For example, see Chapter 7.5.2.

clock design source (as shown in Figure 7.20 on page 273) does not guarantee non-overlapping clocks at the latch/circuit sink points because of distribution skew. That is, logically adjacent circuits may receive clock signals from different clock buffers, which means the respective adjacent clocks may be simultaneously active. Consequently, single phase designs (typically) settle on a separated style.

Storage element support for a single phase separated design is much broader than the master-slave situation. In this style, d-type latches are supported as described in Chapter 5.2.1, in both static and dynamic implementations. Simple d-type static latches as shown in Figure 5.1 can be sequentially used with opposite clock active levels. Similarly, alternating p-type and n-type dynamic latches can also be used, but place restrictions on the incoming data signals (See Section 5.3.2).

Furthermore, a single phase separated latch design also permits the use of single phase master-slave flip-flops. If there is no logic between the latches, then the adjacent d-latches reduce to a master slave flip-flop. However, such a flip-flop cannot be logically placed anywhere. That is, the slave output must ultimately drive a master input, whether through intervening logic or not, as shown in Figure 7.21 on page 274.

Figure 7.22 indicates that either dynamic or static logic may be used in single phase separated phase latch designs. This is true, but the interaction of the static and dynamic logic is restricted, based on the clocking interaction with the logic. That is, there are two possibilities in a single phase system for connecting the clocks and the dynamic circuits. In the first, the clock signal activates (evaluates) the dynamic logic and also opens the subsequent latch. In the second, the clock opens the latch and then activates subsequent dynamic logic.

7.5.3 Clock Activates Logic and Opens Subsequent Latch

In a single phase separated latch design where the clock activates the logic and opens the subsequent latch, static signals driving dynamic logic must be either non-inverting or valid before the clock activates the logic

7.5.4 Clock Opens Latch and Activates Subsequent Logic

In a single phase separated latch design where the clock opens the latch and activates the subsequent latch, static signals must either be either non-inverting or valid before the clock opens the latch.

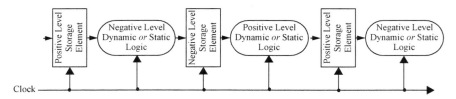

FIGURE 7.23 Single Phase Separated Latch Design: Clock Activates Dynamic Logic and Opens Subsequent Latch

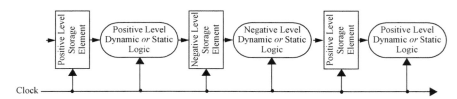

FIGURE 7.24 Single Phase Separated Latch Design: Clock Opens Latch and Activates Subsequent Logic

Advantages	Disadvantages
Simple clocking structure: only a single clock need be generated and distributed.	Limited storage element styles: must use sequentially n-type and p-type latches only.
Good flush-through prevention: single clock allows better skew control/tracking.	*Potentially* limited logic styles: (1) if clock controls logic and the next latch, only dynamic permitted; (2) if clock controls latch and the subsequent logic, both dynamic and static permitted.
Permits use of single phase master-slave design in non-critical areas/paths as long as certain rules are obeyed.	
	Complex interconnection requirements: n-phase logic can only feed n-phase or a p-type latch; p-phase logic can only feed p-phase logic or an n-type latch.
	Poor clock skew tolerance: any introduced skew directly reduces design performance.

TABLE 7-2 Single Phase Separated Latch Advantages and Disadvantages

7.5.5 Single Phase Continuously Latching Design

A continuously latching single phase system stores data at every circuit using alternating nmos/pmos dynamic logic. It is the simplest design possible, but is limited by achievable clock frequency.

Advantages	Disadvantages
Simple clocking structure: only a single clock need be generated and distributed.	Limited latch/logic style: logic and latches must be sequentially nfet and pfet logic stages.
Good flush-through prevention: single clock allows better skew control/tracking.	Complex interconnection requirements: nfet logic/latch can only feed pfet logic/latch.
	Poor clock skew tolerance: any introduced skew directly reduces design performance.
	Does not allow use of any other design style (e.g., single phase master-slave).
	Performance limited to achievable clock speed.

TABLE 7-3 Single Phase Continuously Latching Advantages/Disadvantages

7.6 Multi-Phase Clocking

Multi-phase systems use distinct clock phases generated from multiple main clocks. Logically adjacent latches lie in adjacent clock phases.

Phase generation for multiple phase clocking schemes should be done locally at the end of the tree. This greatly reduces skew by zeroing out most of the distribution differences that would occur between phases if separate distribution trees were attached to each phase. Generating the phases locally at the end of the tree also provided the advantage of only having to distribute one main set of clock tree wires, as opposed to one set of wires per phase.

> **Rule of thumb:** Generating multiple phases locally at the end of a common distribution tree greatly reduces skew,wire,power and noise.

7.6.1 Two Phase Clocking

Two phase clocking systems use two distinct clocks generated from the main clock at the last buffering stage. Logically adjacent latches are separated by half-cycles.

Two Phase Clocking has a high rate of acceptance in the industry. It has good clock skew tolerance, reasonable design simplicity and offers high performance. Adding phases beyond two adds complexity and skew which starts to take away from performance.

Two phase clocking also allows a master-slave system to be used. The two clock phases can be overlapped, underlapped or nominally coincident.

Underlapped two-phase clocking prevents flush-though problems, but can incur delay penalties. Overlapped two-phase clocking helps to prevent delay penalties, but exacerbates flush-through problems. Nominally coincident two-phase clocking is essentially identical to single-phase separated latch design.Hybrid underlapped/overlapped strategies try to solve the flush-through problem and prevent delay penalties.

7.6.2 Four Phase Strategies

Four phase clocking uses four distinct clock phases, each overlapping both the previous and next clock phase, to overcome clock skew. It is very similar to a two phase system.

After considering complexity, wiring, power and skew four phase strategies buy very little for high performance designs, and are not commonly used on high performance microprocessors or ASICS. Just as going beyond two phases is really not required, going beyond four phases buys even less.

7.7 Asynchronous Techniques

Asynchronous techniques avoid the use of all clocks, whether inside a clocked pipeline or on an entire machine basis.

Asynchronous designs do not use a global clock, simplifying global chip routing and eliminating problems due to clock skew. Elimination of the global clock allows an asynchronous chip to achieve near-zero standby power when quiescent. Unfortunately, asynchronous designs are more difficult to implement than the synchronous counterparts. The absence of a global clock requires explicit synchronization between communicating blocks. Glitches which are filtered out by the clock in synchronous designs may cause an asynchronous design to malfunction. Additionally, because asynchronous design styles are not yet widely used, designers cannot find the supporting design tools and methodologies they need. They must manually perform many design tasks that are done automatically for synchronous designers [7.17] .

Sometimes asynchronous interfaces are necessary on chips that are mostly synchronous. For example, a test mode may operate at a slower frequency than functional clocks, a high speed bus may be a higher frequency than the main core clock, or a bus may be slower than the core clock. In these cases techniques are needed to avoid metastability. These techniques involve generating a glitch free interface. Multiple latches in series greatly reduce the chance of a metastable condition occurring; use of metastable hardened latches also helps. Since the metastability equation is a function of capture and sending frequencies, these factors can also be used to help reduce the chance of metastability. Such a condition could occur on RAMs. (e.g., reading and writing of the same word). A fix for this condition is to insure that the logic interface to insure a read and write of the same word do not occur.

Other Asynchronous techniques include self-resetting Strategies and self-timed strategies. Self-resetting approaches send information in pulses, but are prone to sink circuits missing information.Self-timed approaches require sink circuits to confirm receipt of information, but incur loop delays and additional wiring complexity.

7.8 Summary

Clock Generation schemes were described for both on-chip and off-chip solutions, Jitter and skew, which limit clock performance, were described, and their underlying causes were discusses. Two-phase and single-phase clock systems were then compared.

The challenges of clock distribution design continue to increase. A variety of alternatives including package wiring, and asynchronous logic can reduce the need for accurate on-chip clock distributions, however the increased cost and complexity of these methods will limit their widespread use as long as strategies and technologies to design efficient synchronous clocking continue to succeed. The sophisticated interconnect modeling techniques (previously used mainly for chip-to-chip interconnects) and design tools required to design clock networks are also becoming important for other more numerous medium and long on-chip wiring. On-chip reference planes simplify extraction, but are not presently needed to control on-chip transmission-line effect.

References

[7.1]   L. A. Glasser and D. W. Dobberpuhl, *The Design and Analysis of VLSI Circuits*, Addison Wesley Publishing Company Inc., 1985, p. 345.

[7.2]   T. Fischer and D. Leibholz, "Design Tradeoffs in Stall-Control Circuits for 600MHz Instruction Queues," *ISSCC Digest of Technical Papers*, Feb. 1998, p. 232.

[7.3]   N. Weste and K. Eshraghian, *Principles of CMOS VLSI Design*, Second Edition, Addison Wesley Publishing Company Inc., 1995, p. 417.

[7.4]   A. Deutsch *et al*, "When are Transmission-Line Effects Important for On Chip Interconnections?," *IEEE Trans. Microw. Theory Tech.* (USA) Vol. 45, No. 10, pt. 2, Oct. 1997, pp. 1836-46.

[7.5]   B. Gieseke *et al*, "A 600 MHz Superscalar RISC Microprocessor With Out-of-Order Execution," *ISSCC Tech. Dig.* Feb. 1997, pp. 176-7.

[7.6]   C. Webb *et al* "A 400MHz S/390 Microprocessor," *IEEE JSSC*, vol. 32, no. 11, 1997, pp. 1665-1675.

[7.7]   P.J. Restle, K.A. Jenkins, A. Deutsch and P.W. Cook, "Measurement and Modeling of On-Chip Transmission-Line Effects in a 400 MHz Microprocessor," *IEEE JSSC*, Vol. 33 No. 4, Apr. 1998, pp. 662-665.

[7.8]   G. Geannopoulos, X. Dai, "An Adaptive Digital Deskewing Circuit for Clock Distribution Networks," *IEEE SSCC Tech. Dig.*, Feb. 1998, p. 400.

[7.9]   D. Edelstein *et al*, "Full Copper Wiring in a Sub-0.25 lm CMOS ULSI Technology," *IEEE IEDM Tech. Dig.*, Dec. 1997, pp. 773-6.

[7.10]   J.A. Kash and J.C. Tsang, "Dynamic Internal Testing of CMOS Circuits Using Hot Luminescence," *IEEE Electron Device Lett.*, vol. 18, 1997, pp. 330-332.

[7.11]   E. G. Friedman (ed.), *Clock Distribution Networks in VLSI Circuits and Systems*, Piscataway, NJ, IEEE Press, 1995.

[7.12]   K. M. Carrig *et al*, "A Methodology and Apparatus for Making a Skew-Controlled Signal Distribution Network," U.S. patent no. 5,339,253, filed June 14, 1991, issued Aug. 16, 1994.

[7.13]   K. M. Carrig *et al*, "A Clock Methodology for High-Performance Microprocessors," in *Proc. IEEE Custom Integrated Circuits Conference*, May 1997, pp. 119-122.

[7.14]   K. M. Carrig *et al*, "A New Direction in ASIC High-Performance Clock Methodology," in *Proc. IEEE Custom Integrated Circuits Conference*, May 1998.

[7.15]   A. M. Rincon, M. Trick, and T. Guzowski, "A Proven Methodology for Designing One-Million-Gate ASICs," *Proc. IEEE Custom Integrated Circuits Conference*, May 1996, pp. 45-52.

[7.16]   M. Shoji, "Elimination of Process-Dependent Clock Skew in CMOS VLSI," *IEEE J. Solid-State Circuits*, vol. SC-21, no. 5, Oct. 1986, pp. 875-880.

[7.17]   A.Marshall *et al*, "Designing an Asynchronous Communications Chip," *IEEE Design & Test of Computers,* vol. 11,no. 2,Summer 1994, pp. 8-21.

CHAPTER 8

Slack Borrowing and Time Stealing

8.1 Introduction

With any circuit, clocking, and latching selection, the concept of how to fit more logic within a path between latches than is readily available always becomes an issue. That is, inevitably a logical pipeline partition will require more time than is available, for example, more than a full-cycle time in a master-slave system or a half-cycle in a two-phase separated-latch system. Depending on the circuit style, the latching structure, and the clocking strategy, obtaining this time can be classified as one of two categories, *slack borrowing* and *time stealing* (also commonly referred to as *cycle stealing*).

Slack borrowing refers to the case where a logical partition utilizes time *left over* (*slack* time) by the *previous* partition. Additionally, slack borrowing requires this utilization of additional time to be automatic. Consequently, the surrendering of time is considered voluntary by the previous partition. A key point to slack borrowing is that the slack utilization is performed without circuitry or clock arrival time adjustments.

Time stealing, on the other hand, occurs where a logical partition utilizes a portion of the time *allotted* to the *next* partition. The additional time is stolen from the subsequent partition's allotment. This necessitates forcible removal of time, so time stealing is considered non-automatic and involuntarily surrendered. In general, the additional time is obtained by adjusting the clock arrival time(s).

In essence, both slack borrowing and time stealing permit pipeline partitions to use more than the allotted time between successive registers and still maintain the overall machine cycle time. That is, if the time used for logic evaluation between two successive registers exceeds one cycle time and the overall machine still works at speed, then that excess time requirement has either been borrowed from the preceding pipeline stage or stolen from the subsequent pipeline stage.

In this chapter we focus on slack borrowing and time stealing in a *two-phase* clocking system, essentially independent of circuit and latching style selection.[1] The discussions cover both static and dynamic logic circuit styles using transparent and master-slave latches, using many examples to clarify the concepts.

8.2 Slack Borrowing

Slack borrowing permits logic to *automatically* use slack time from a *previous* cycle.

Slack borrowing refers to the case where a logical partition utilizes time *left over* *(slack)* by the *previous* partition. By definition, this additional time is automatic and is consequently considered as voluntarily surrendered. Also by definition, slack borrowing is performed without circuitry and/or clock arrival time adjustments.

The very nature of slack borrowing precludes the use of both edge-triggered logic and edge-triggered latching. That is, edge-triggered structures require a clock edge to begin evaluation, thus necessitating the movement of a clock edge in order to obtain more evaluation time. Thus, to utilize *previous* partition time, the logical evaluation must begin earlier, which means that an earlier clock signal must be used. Since, however, the clock signal is (generally) routed to produced a rigidly defined arrival time (see Chapter 7), an earlier clock signal may not be available. Additionally, because using an earlier clock signal is by definition a clock arrival time adjustment at the circuit, it violates the definition of slack borrowing. Consequently, although it is feasible to utilize left-over time in the previous partition by using an earlier clock to begin evaluation, it is not slack borrowing, but is an extended case of time stealing. As a result, slack borrowing precludes the use of edge-triggered logic families, such as dynamic logic, and latches, such as master-slave flip-flops.

---

1. Although some have at least *proposed* multi-phase clocking (see, for example, [8.1]) this is not common in industry. Rather, two-phase clocking presently dominates (see, for example, [8.2] and [8.3]).

Slack borrowing is, then, a technique used with static logic and in systems utilizing non-edge-triggered (transparent) latches. Since they require no clocks for operation, static logic circuits begin evaluating as soon as the data inputs arrive, producing a *data-gated* operation.[2] Non-edge triggered (transparent) latches propagate input data to the output when the clock signal is active.[3] Consequently, as long as the clock signal is active such a latch is also *data-gated*. The interaction of these two is slack borrowing.

Therefore, because slack borrowing is concerned only with static logic and transparent latches, there is but one selection for clocking with one selection for latching: two phase clocking with separated latches.

> **Rule of thumb:** Slack borrowing is only used with non-edge-triggered circuits and latches, and is ideally suited for static logic in a two-phase clocking system utilizing separated latch design techniques.

Two-phase clocking and separation of the latches results in two basic divisions in slack borrowing: *cycle* and *phase*.

Cycle slack borrowing permits logic pipelines between cycle boundaries to use more than one cycle time and still fit within a single cycle clock boundary *while* maintaining the overall machine cycle time. That is, if the time used for logic evaluation in a cycle boundary *exceeds* one cycle and the machine still works at speed, then slack time has been borrowed from preceding phase(s).

Similarly, ***phase slack borrowing*** allows logic in a particular clock phase (typically one-half of a cycle) to use more time than is readily available and *still* maintain the overall machine cycle time. That is, if the time used for logic evaluation in a particular clock phase *exceeds* the clock phase time and the machine still works at speed, then slack time has been borrowed from preceding cycle(s).

A primary ramification of slack borrowing is ***phase partitioning***. It refers to the optimal positioning of latches for performance and area.

To fully understand slack borrowing and phase partitioning, we first consider a symmetric 50% duty cycle two-phase system, then extend the ideas into an asymmetric two-phase system. These are followed by several examples.

---

2. See, for example, Section 2.2.1.
3. See Section 5.2.1 on page 177.

8.2.1 Slack Borrowing in Symmetric 50% Duty Cycle 2-Phase Systems

Slack borrowing in symmetric 50% duty cycle two-phase systems is optimal.

A symmetric 50% duty cycle two-phase system is one in which the centrally generated and distributed system clock is split into two identical but logically inverted clocks.[4] Figure 8.1 shows a timing diagram of the clocks, illustrating that each is active (high) for one-half of the cycle and inactive (low) for the other one-half of the cycle. Note in Figure 8.1 that the individual clock edges are pictured as aligned with the system clock, which is an idealized system for conceptual understanding. In a real application, there would be some delay to generate these clocks from the system clock and some amount of clock skew that would preclude all clock edges from an ideal alignment.

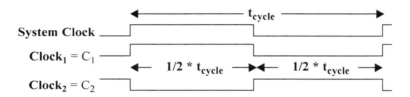

FIGURE 8.1 Symmetric 50% Duty Cycle 2-Phase Clock System

Typically, the clock signal denoted "Clock$_1$" (or "C$_1$") is defined as the "cycle boundary clock" as it rises (and falls) with the system clock, while "Clock$_2$" (or "C$_2$") is the "mid-cycle boundary clock." The definitions are essentially arbitrary, but differentiate between cycle slack borrowing and phase slack borrowing. That is, swapping the definitions is effectively meaningless.[5]

In a chip application utilizing slack borrowing, the clocks of Figure 8.1 are used to control a two-phase separated latch system as shown in Figure 8.2 on page 289. Note that the clocks control transparent latch elements alternately, creating a pipeline system.[8.5]

In light of Figure 8.1 and Figure 8.2, it is important to make several clarifying definitions before discussing slack borrowing.

---

4. See Section 7.6.1.

5. This is true unless special extensions are made to the clock definitions for testability purposes.[8.4]

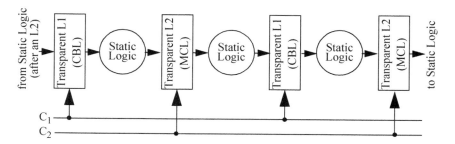

FIGURE 8.2 2-Phase Separated Latch System using Static Logic

- **System Clock:** The centrally routed/distributed clock in a two-phase system.

- **Cycle Boundary Clock:** The clock signal whose falling edge signifies the end of the system cycle. Equivalently, the clock signal that is active in the second half-cycle of the two-phase system. Note that this definition is actually arbitrary and can be swapped with that of the *mid-cycle boundary clock*.

- **Mid-Cycle Boundary Clock:** The clock signal whose falling edge signifies the end of the first hale of the system cycle. Equivalently, the clock signal that is active in the first half-cycle of the two-phase system. Note that this definition is actually arbitrary and can be swapped with that of the *cycle boundary clock*.

- **Transparent Latch:** A storage element that passes data when the clock is high and stores (the last) data when the clock is low.

- C_1: The clock signal that is *inactive* in the *first* half of the system clock cycle and *active* in the *second*; typically called the "cycle boundary clock."

- C_2: The clock signal that is *active* in the *first* half of the system clock cycle and *inactive* in the *second*; typically called the "mid-cycle boundary clock."

- **L1:** A latch whose clock signal is connected to the C_1 Clock.

- **L2:** A latch whose clock signal is connected to the C_2 Clock.

- **CBL:** Cycle Boundary Latch, or the latch whose clock signal is connected to the cycle boundary clock; equivalent to an L1 if C_1 is the cycle boundary clock.

- **MCL:** Mid-Cycle Latch, or the latch whose clock signal is connected to the mid-cycle boundary clock; equivalent to an L2 if C_2 is the mid-cycle boundary clock.

- **Static Logic:** Logic which does not require a clock for operation and is not self-timed. See Chapter 2 for sample logic families and details.

With these definitions an updated diagram (Figure 8.3) that presents the basics for slack borrowing can be created which combines the clocking diagram of Figure 8.1 and the system diagram of Figure 8.2.

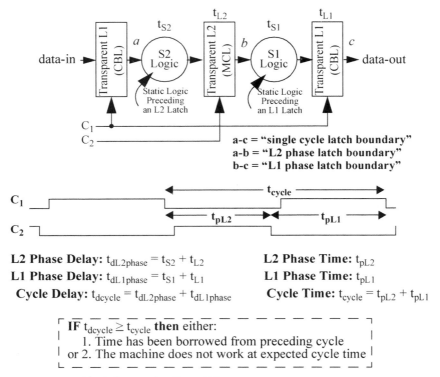

L2 Phase Delay: $t_{dL2phase} = t_{S2} + t_{L2}$ **L2 Phase Time:** t_{pL2}
L1 Phase Delay: $t_{dL1phase} = t_{S1} + t_{L1}$ **L1 Phase Time:** t_{pL1}
   Cycle Delay: $t_{dcycle} = t_{dL2phase} + t_{dL1phase}$ **Cycle Time:** $t_{cycle} = t_{pL2} + t_{pL1}$

> **IF** $t_{dcycle} \geq t_{cycle}$ **then** either:
>   1. Time has been borrowed from preceding cycle
> or 2. The machine does not work at expected cycle time

FIGURE 8.3 Slack Borrowing Base Diagram

Note in Figure 8.3 that the clock signals (C_1 and C_2) are shown with non-overlapping active (high) levels. Denoted as an underlapped system,[6] this prevents the possibility of a flushing system where two successive latches (CBL and MCL) are data-gated (clocks active) simultaneously, which would cause the overall pipeline system to lose the required logic state. Such a problem is deemed an "early mode" or "fast path" problem. Section 7.6 covers this problem in more detail.

---

6. See Section 7.6.

Because *cycle* slack borrowing by definition means that a logical pipeline stage between any successive cycle boundary latches may actually take longer than a cycle to evaluate, it is important to distinguish between cycle *time* and cycle *delay*.

- **Cycle Time** is the time between successive C_1 clock falling edges and is denoted "t_{cycle}." Equivalently, it is the inverse system of the clock frequency (f_{clock}) or

$$t_{cycle} = 1 / f_{clock}. \qquad (8.1)$$

- **Cycle Delay** is denoted "t_{dcycle}" and refers to the delay used by the logic within a single cycle latch boundary. It may be *greater* or *less* than a cycle time.

Furthermore, we can divide the single cycle latch boundary in Figure 8.3 (*a* to *c*) into the "L2 phase latch boundary" (*a* to *b*) and the "L1 phase latch boundary" (*b* to *c*) and make three more definitions.

- **Single Cycle Latch Boundary** is the boundary from the output of the *previous* CBL to the output of the *next* CBL. It contains the logic following the previous CBL, the MCL, the logic following the MCL, and the next CBL.
- **L2 Phase Latch Boundary** is the boundary from the output of the *previous* CBL to the output of the MCL. It contains all the logic after the CBL and the next MCL.
- **L1 Phase Latch Boundary** is the boundary from the output of the MCL to the output of the *next* CBL. It contains the logic after the MCL and the next CBL.

Similar to *cycle* slack borrowing, *phase* slack borrowing by definition means that logic in a phase latch boundary may actually take longer than one-half a machine cycle time. It is also important, then, to further delineate phase *time* and phase *delay*.

- **L2 Phase Time** is the time from the falling edge of the C_1 clock to the falling edge of the C_2 clock. Denoted "t_{pL2}" for symmetric 50% duty cycle clocks its value is one-half the cycle time, or

$$t_{pL2} = 0.5 * t_{cycle}. \qquad (8.2)$$

- **L2 Phase Delay** is denoted "$t_{dL2phase}$" and refers to the delay used by the logic within the L2 phase latch boundary (t_{S2}) *plus* the time to propagate data through the L2 latch (t_{L2}) or

$$t_{dL2phase} = t_{S2} + t_{L2}. \qquad (8.3)$$

Note that the L2 phase delay may also be *greater* or *less* than the L2 phase time.

- **L1 Phase Time** is the time from the falling edge of the C_2 clock to the falling edge of the C_1 clock and is denoted "t_{pL1}." For symmetric 50% duty cycle clocks, its value is one-half the cycle time, or

$$t_{pL1} = 0.5*t_{cycle}. \tag{8.4}$$

- **L1 Phase Delay** is denoted $t_{dL1phase}$ and refers to the delay used by the logic within the L1 phase latch boundary (t_{S1}) *plus* the time to propagate data through the L1 latch (t_{L1}) or

$$t_{dL1phase} = t_{S1} + t_{L1}. \tag{8.5}$$

Note that the L1 phase delay may be *greater* or *less* than the L1 phase time.

We can further define cycle delay as the sum of the phase delays, or

$$t_{dcycle} = t_{dL2phase} + t_{dL1phase}. \tag{8.6}$$

Slack borrowing can be summarized rather simply: if in a system of the type shown in Figure 8.2, a given pipeline latch/phase boundary *delay* exceeds the latch/phase *time*,

$$t_{dcycle} > t_{cycle} \quad or \quad t_{dLxphase} > t_{pLx} \tag{8.7}$$

and the machine still works at the desired frequency, or

$$f_{cycle} = f_{desired} \tag{8.8}$$

then slack borrowing has occurred.

To define the parameters and rules of slack borrowing, consider Figure 8.4 on page 293, which is a detailed example of slack borrowing. Comparing this figure to Figure 8.3 on page 290, two major differences should be noted. First, the static logic blocks (S2 and S1) have been split into two sections each ("a" and "b"). Second, the clock diagram indicates two new time notations - "t_{gap12}" and "t_{gap21}" - between the falling and rising edges of the clock signals. These "gaps" represent the time introduced by clock routing and generation as well as any desired underlapping designed into the clock system. Additionally, no set-up and hold times are noted in the timing diagrams, as is required of any latch.[7] This omission also reduces the figure and discussion complexity significantly with little impact on content.

---

7. For example, see Section 5.2.3.

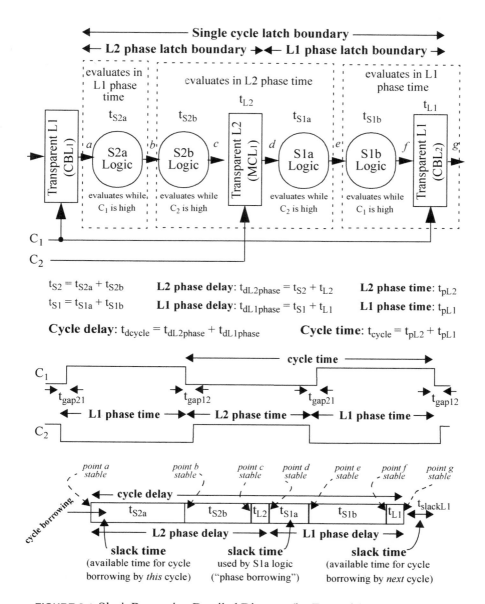

FIGURE 8.4 Slack Borrowing Detailed Diagram (by Example)

Referring to Figure 8.4, observe that point a is stable prior to the close of latch of CBL_1 as clock C_1 is high. This means that the data input to CBL_1 arrived at the time a was stable less the latch propagation time (t_{L1}). But, because CBL_1 is a *transparent* latch, the data could have arrived up to the end of the L1 phase time. This, then, indicates that the logic in the previous L1 phase time did not utilize all of its available evaluation time, which results in left over time, or slack. This time is *borrowed* by the static logic section S2a. That is, because S2a is by definition *static* logic, it begins logical evaluation when the input (a) is stable, or in the *previous* L1 phase time (while C_1 is high), even though it resides in the L2 phase latch boundary.

Thus, S2a produces the result b in a time delay of t_{S2a}, or b becomes stable at the end of the L1 phase time (fall of C_1). But note that this is the required time where the data input to CBL_1 must be stable. Consequently, the S2a logic portion *appears* to be on the *input* side of CBL_1, although it *physically* resides on the *output* side. This means that S2a *could* physically reside on the input side if desired with no effect on the logical timing.

When b stabilizes, S2b evaluates in time t_{S2b} to produce c while clock C_2 is active (high) or during the L2 phase time. Note that c is the data input to MCL_1 and is required to be stable at the fall of the C_2 clock, but has arrived (in this example) well before that time. Since, then, C_2 is active, MCL_1 produces its output d after the latch propagation time (t_{L2}).

Observe that point d is stable prior to the close of latch of MCL_1 as clock C_2 is high. Because MCL_1 is a *transparent* latch, the data could have arrived up to the end of the L2 phase time (fall of C_2). Thus, the S2b logic in the L2 phase time did not utilize all of its available evaluation time, which results in left over time, or slack. This time is *borrowed* by the static logic section S1a. That is, because S1a is by definition *static* logic, it begins logical evaluation when the input (d) is stable, or in the *previous* L1 phase time (while C_1 is high), even though it resides in the L2 phase latch boundary.

S1a evaluates d in time t_{S2a} to produce e at the falling edge of clock C_2, or at the end of the L2 phase time. But, this is the required time where the data input to MCL_1 must be stable. Consequently, the S1a logic portion *appears* to be on the *input* side of MCL_1, although it *physically* resides on the *output* side. This means that S1a *could* physically reside on the input side if desired with no effect on the logical timing.

Furthermore, when e stabilizes, S1b evaluates in time t_{S1b} to produce f while clock C_1 is active (high) or during the L1 phase time. Here, e is the data input to CBL_2 and is required to be stable at the fall of the C_1 clock. Note that it has arrived (in this example) well before that requirement (similar to point c). So, because C_1 is active, CBL_2 produces its output g after the latch propagation time (t_{L1}).

We can extract timing information from Figure 8.4. For example, using equation (8.3), the L2 phase delay is the sum of all the events in the L2 phase latch boundary, or from *a* to *d* in Figure 8.4, or

$$t_{dL2phase} = t_{S2a} + t_{S2b} + t_{L2} \tag{8.9}$$

which is the same as equation (8.3) except that S2 has been split into S2a and S2b. Similarly, from equation (8.5), the L1 phase delay is the sum of all the events in the L1 phase latch boundary, or from *d* to *g* in Figure 8.4, or

$$t_{dL1phase} = t_{S1a} + t_{S1b} + t_{L1} \tag{8.10}$$

which is the same as equation (8.5) except that S1 has been split into S1a and S1b. Using equation (8.6), this means the cycle delay is

$$t_{dcycle} = t_{dL2phase} + t_{dL1phase} = t_{S2a} + t_{S2b} + t_{L2} + t_{S1a} + t_{S1b} + t_{L1}. \tag{8.11}$$

Additionally, equation (8.2) states that the L2 phase time is equal to one-half of the cycle time. However, it is defined as the time from C_1 falling edge to the time C_2 falling edge. Therefore, it can be alternatively expressed as in terms of the sum of the times of events that occur within that span, or

$$t_{pL2} = 0.5*t_{cycle} = t_{S2b} + t_{L2} + t_{S1a}. \tag{8.12}$$

Similarly, the L1 phase time is also equal to one-half of the cycle time, but is defined as the time from C_2 falling edge to the time C_1 falling edge. It, too, can be alternatively expressed as in terms of the sum of the times of events that occur within that span, or

$$t_{pL1} = 0.5*t_{cycle} = t_{S1b} + t_{L1} + t_{slackL1}. \tag{8.13}$$

The cycle time can be re-written as

$$t_{cycle} = 0.5*t_{cycle} + 0.5*t_{cycle} = t_{pL2} + t_{pL1} = t_{S2b} + t_{L2} + t_{S1a} + t_{S1b} + t_{L1} + t_{slackL1} \tag{8.14}$$

Subtracting equation (8.14) from equation (8.11) provides

$$t_{dcycle} - t_{cycle} = t_{S2a} - t_{slackL1}. \tag{8.15}$$

Equation (8.15) provides a simple way to look at *cycle* slack borrowing. Consider the following three cases concerning the right-hand side of the equation.

1. $t_{S2a} = t_{slackL1}$: The cycle delay is *equal to* the cycle time.

When the cycle delay is equal to the cycle time, no additional time is required for the single cycle latch boundary to complete its logical evaluation. However, this does *not* mean that slack borrowing has not occurred. Referring to Figure 8.4, remember that both S2a and S1a borrowed slack time, but time was left over by the previous cycle. Note that this time is now available for borrowing by the *next* cycle. Consequently, *this* cycle boundary has passed the slack time from the *previous* cycle to the *next* cycle, which is the first rule of slack borrowing:

Slack Borrowing Rule: A cycle may pass available slack time from the previous cycle across its boundary to the next cycle.

2. $t_{S2a} > t_{slackL1}$: The cycle delay is *greater than* the cycle time.

When the cycle delay is greater than the cycle time, slack borrowing must have occurred, or else the machine would not operate at the desired frequency, as the single cycle latch boundary needed additional evaluation time to complete. Note that this does not prevent passing *previous*ly available slack time to the *next* cycle. That is, a cycle may either use all the available slack time from the previous cycle ($t_{slackL1}$ is zero) or reduce in the amount of passed slack time ($t_{slackL1}$ is greater than zero, but less than t_{S2a}) which is another rule:

Slack Borrowing Rule: A cycle may eliminate or decrease the amount of slack time passed across its boundary to the next cycle.

3. $t_{S2a} < t_{slackL1}$: The cycle delay is *less than* the cycle time.

When the cycle delay is less than the cycle time, the single cycle latch boundary has slack time available for borrowing by the next cycle. Note, again, that this does not mean that slack borrowing has not occurred. Rather, it means the cycle increases the amount of slack time borrowed from the *previous* cycle which is made available for the *next* cycle:

Slack Borrowing Rule: A cycle that does not use a full cycle time generates additional slack time, increasing the amount of slack available for the next cycle.

Furthermore, consider the limit on the time the S2a logic may borrow. That is, S2a begins starts its evaluation when a is stable, which can be any time during the L1 phase time. Therefore, the evaluation time for S2a can be up to the limit of where a is stable before the fall of C_1. Note, then, the earliest a can be stable occurs just after the rise of the C_1 clock (as CBL_1 must be open). This means the input to CBL_1 must have awaited the opening the rise of C_1, or that a became stable from the rising edge of C_1.

If we define the time CBL_1 requires to launch the data from the rising edge of C_1 to its output as "$t_{launchL1}$" then the maximum time S2a can use for evaluation is:

$$\text{maximum}\{t_{S2a}\} = t_{pL1} - t_{gap12} - t_{launchL1} = 0.5*t_{cycle} - t_{gap12} - t_{launchL1}. \quad \text{(8.16)}$$

Neglecting the gap time and the launch time,[8] this reduces to

$$\text{maximum}\{t_{S2a}\} \cong 0.5*t_{cycle}. \qquad \textbf{(8.17)}$$

Slack Borrowing Rule: A cycle may borrow a maximum of one-half of the cycle time, neglecting the latch launch and clock edge gap times.

Thus, because slack time can accumulate across multiple cycles, passing from one cycle to the next, the maximum amount of time that can be accumulated and passed is also one-half a cycle time.

Slack Borrowing Rule: The cumulative slack that can be passed across a cycle(s) can be a maximum of one-half of the cycle time, neglecting the latch launch and clock edge gap times.

Equation (8.17) can be extended to describe the maximum cycle delay by noting that is simply the sum of the cycle time plus the amount of borrowed time, or

$$\text{maximum}\{t_{dcycle}\} \cong 1.5*t_{cycle}. \qquad \textbf{(8.18)}$$

Slack Borrowing Rule: A cycle delay may be a maximum of one-and-one-half times the cycle time, neglecting the latch launch and clock edge gap times.

Expanding the scope of Figure 8.4 to a chip, suppose a pipelined machine has a number of pipeline stages n, each requiring one cycle time.[8.5] Since the *overall* machine cannot use more time than is available, the total delay of the machine is

$$t_{total-delay} = n*t_{cycle}. \qquad \textbf{(8.19)}$$

Otherwise, the desired frequency will violate equation (8.8). This means that the sum of all the cycle delays cannot exceed the total delay of the n-stage pipeline machine.

Slack Borrowing Rule: The sum of all cycle delays across an n-stage pipeline machine can be a maximum of n times the cycle time.

For *phase* slack borrowing, we can derive another equation similar to equation (8.15). That is, we can subtract the *phase time* from the *phase delay* for both the L1 and L2 phases. For the L2 phase, we subtract equation (8.12) from equation (8.9):

$$t_{dL2phase} - t_{pL2} = t_{S2a} - t_{S1a}. \qquad \textbf{(8.20)}$$

Similarly, we subtract equation (8.13) from equation (8.10) for the L1 phase:

---

8. The gap time is *typically* inconsequential when compare to the chip cycle time. Additionally, *typically* the latch launch time (t_{Launch}) is approximately equal to the data delay of the latch (t_{Lx}) (see Section 5.2.3) and is also *typically* much smaller than the gap time. Thus, neglecting these times introduces only a small error in the computations.

$$t_{dL1phase} - t_{pL1} = t_{S1a} - t_{slackL1}. \tag{8.21}$$

Concentrating on the L2 phase, consider the right hand side of equation (8.20). As in the cycle slack borrowing case, the right-hand side of this equation presents three possible cases for the L2 phase:

1. $t_{S2a} = t_{S1a}$: The phase delay is *equal to* the phase time.

   When the L2 phase delay is equal to the phase time, no additional time is required for the L2 phase latch boundary to complete its logical evaluation, which does *not* mean that slack borrowing has not occurred. Note that S2a borrowed slack time, but an equivalent amount of time was left over and used by S1a. Consequently, the L2 phase has passed the slack time from the *previous* L1 phase to the *next*:

   Slack Borrowing Rule: A phase may pass available slack time from the previous phase across its boundary to the next phase.

2. $t_{S2a} > t_{S1a}$: The phase delay is *greater than* the phase time.

   When the phase delay is greater than the phase time, slack borrowing must have occurred (else the machine would not operate at the desired frequency) as the L2 phase latch boundary needed additional evaluation time to complete. Note, though, that this does not prevent passing *previously* available phase slack time to the *next* phase. That is, a phase may either use all the available slack time from the previous phase (t_{S1a} is zero) or reduce in the amount of passed slack time (t_{S1a} is greater than zero, but less than t_{S2a}).

   Slack Borrowing Rule: A phase may eliminate or decrease the amount of slack time passed across its boundary to the next phase.

3. $t_{S2a} < t_{S1a}$: The phase delay is *less than* the phase time.

   When the phase delay is less than the phase time, the phase latch boundary has slack time available for borrowing by the next phase, which, again, does not mean that slack borrowing has not occurred. Rather, it means the phase increases the amount of slack time borrowed from the *previous* phase, for use by the *next* phase:

   Slack Borrowing Rule: A phase that does not use a full phase time generates additional slack time, increasing the amount of slack available for the next phase.

A similar derivation can be made for the L1 phase using equation (8.21), but would provide the same rules as for the L2.

Furthermore, for *cycle* slack borrowing, equations (8.16) and (8.17) illustrated the maximum amount of delay S2a could borrow from the L1 phase latch boundary. Because these equations were concerned with the amount of time the L2 phase latch boundary could borrow from the L1 phase, they are identically applicable for *phase* slack borrowing as well as cycle slack borrowing:

Slack Borrowing Rule: A phase may borrow a maximum of one-half of the cycle time, neglecting the latch launch and clock edge gap times.

Thus, because slack time can accumulate across multiple phases, passing from one phase to the next, the maximum amount of time that can be accumulated and passed is also one-half a cycle time.

Slack Borrowing Rule: The cumulative slack that can be passed across a phase(s) can be a maximum of one-half of the cycle time, neglecting the latch launch and clock edge gap times.

Additionally, equation (8.17) can be extended to describe the maximum phase delay similar to equation (8.18) by noting the maximum phase delay is equal to the phase time plus the maximum borrowed time:

$$\text{maximum}\{t_{dLxphase}\} \cong 1 * t_{cycle}. \tag{8.22}$$

Slack Borrowing Rule: A phase delay may be a maximum of the cycle time, neglecting launch time of the latch and the gap time between clock edges.

In addition to the preceding rules governing the passing of slack time and its limitations, several other boundaries exist concerning slack borrowing. Consider the timing diagram of Figure 8.5 on page 300, which is applicable to the system diagram of Figure 8.4 on page 293. Note that logic block S1b has been removed, or, equivalently, its delay (t_{S1b}) is zero.

Without block S1b, note that points e and f are identical. Note in Figure 8.5 that f arrives prior to the rise of the C_1 clock, which also indicates the opening of latch CBL_2. Consequently, point g must await the opening of the latch to stabilize, or the launch delay of the latch ($t_{launchL2}$) which is approximately equal to the data delay of the latch (t_{L2}).[9] As a result, in the L1 phase delay a dead time occurs, during which no

logical execution can occur. That is, because the logic after CBL_2 does not have a stable input (g), it cannot evaluate. Note that an identical situation would occur if point c were stable prior to the rising edge of the C_2 clock for MCL_1.

Slack Borrowing Rule: If the data to any latch arrives when the latch is closed, a dead time penalty is paid equal to the amount of time the data arrives prior to the latch opening (rise of the clock) assuming the launch delay of the latch is approximately equal to the data delay.

Furthermore, consider the effects of the inherent clock edge inaccuracy. The time any given clock edge arrives is subject to movement from process, circuit, and environ-

---

9. See Section 5.2.3.

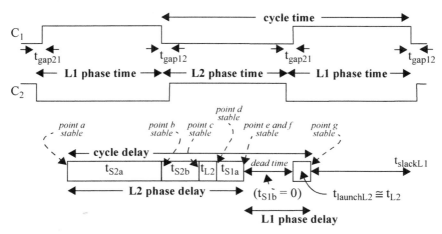

FIGURE 8.5 Slack Borrowing Timing Diagram with S1b Removed (by Example)

ment variability. As described in Chapter 1, process technology is in many cases within 10% on a *single* given parameter at best.[10] This translates directly into edge inaccuracy in the clock circuitry. Additional variability comes from the clock generation circuitry, e.g., a Phase-Locked-Loop (PLL), and its inability to completely mimic the system-level clock.[11] Also, environmental issues such as chip and board temperature gradients contribute to the clock edge inaccuracy.[12] These effects create an uncertainty in the arrival time of the clock edge, both earlier and later in time, and is generally lumped into two categories: jitter and skew. [13]

Referring back to Figure 8.5, because the clock edge launches the output of CBL_2, the its arrival time inaccuracy affects the known stability time of *g*. Thus, in addition to the dead time penalty assessed (because points *e* and *f* arrived before CBL_2 was open), there is an additional time penalty associated with the inaccuracy of the clock edge. This is called "latch relaunch penalty" and is equal to the difference between the latch launch delay and data delay plus the clock skew and clock jitter amount:

$$t_{relaunch-penalty} = t_{launchLx} - t_{Lx} + t_{clock-jitter} + t_{clock-skew} \qquad (8.23)$$

---

10. For example, see Section 1.3.2.

11. That is, PLLs lock on the incoming frequency, but the increments of the lock-in are not infinite. Thus, they add an amount of additional uncertainty in the clock edge inaccuracy.

12. See Sections 7.2.1 and 7.2.2.

13. See Section 7.2.

If we again assume that the launch and data delays of the latches are nearly equal, this reduces to

$$t_{\text{relaunch-penalty}} \cong t_{\text{clock-jitter}} + t_{\text{clock-skew}}. \qquad (8.24)$$

Note that this occurs at any latch boundary (CBL or MCL). Furthermore, this effect occurs anytime a latch output stability is determined by the clock signal, even if the dead time is zero. Additionally, if the input to a latch arrives at the clock edge boundary within the jitter and skew time of the clock edge itself, then a latch relaunch penalty must be assessed as the clock edge may occur later in time.

> **Slack Borrowing Rule:** A latch relaunch penalty equal to the amount of clock edge inaccuracy must be assessed if the input to a latch arrives within the edge inaccuracy time of the clock edge, assuming the launch delay of the latch is approximately equal to the data delay.

Consequently, it is important to note that if the data arrives while the clock is active (high) at any MCL or CBL circuit, then no dead time or latch relaunch penalty can occur. Thus, it is important to position the latches such that the inputs always arrive while the clock is active.[14] This gives rise to an important ramification of slack borrowing known as *phase partitioning*.

8.2.2 Phase Partitioning in Symmetric 50% Duty Cycle 2-Phase Systems

Phase partitioning allows latches to move across logic circuitry within a phase time to prevent relaunch penalties and/or save time, area, and power.

Phase partitioning is the ability to optimally position latches, both MCLs and CBLs, in time to prevent relaunch and dead time penalties. That is, if the inputs to latches always arrive while the clocks are active (or, alternatively, while the latches are open), then each latch is data-gated (transparent).

For example, note in Figure 8.4 on page 293 that MCL_1 is positioned between the S2b and S1a logic segments. Because the MCL_1 is transparent while C_2 is high, it may, then, be moved anywhere inside this time, reapportioning the logic between S2b and S1a. This can be extended to any MCL or CBL to create the first rule of phase partitioning:

> **Phase Partitioning Rule:** A latch may be moved anywhere inside a phase time, or, equivalently, logic may be moved across a latch boundary within a phase time.

---

14. Note that in complex applications (such as super-scalar microprocessors) it is essentially impossible to force *all* latch inputs across *all* latch boundaries in the machine pipeline to arrive while the clocks are high. Consequently, some number of relaunch penalties are inevitable.

Obviously MCL_1 cannot move past the time where the C_2 clock is high in either direction. Thus, the maximum MCL movement time is the L2 phase time less the gap times, or

$$\text{maximum MCL movement} = t_{pL2} - t_{gap12} - t_{gap12}. \tag{8.25}$$

If we ignore the gap times and use equation (8.2) for the phase time, this reduces to

$$\text{maximum MCL movement} \cong t_{pL2} = 0.5*t_{cycle}. \tag{8.26}$$

Equation (8.26) can be easily be extended to any CBL as well:

$$\text{maximum CBL movement} \cong t_{pL1} = 0.5*t_{cycle}. \tag{8.27}$$

Equations (8.26) and (8.27) describe a second rule of phase partitioning:

Phase Partitioning Rule: The maximum latch movement is one-half of a cycle time, neglecting the gaps times.

The ability to move latches across logic inside a phase time, then, provides a "soft" latch boundary. That is, the latch location is not important if the data arrives while the latch clock is high. However, latch movement is not solely constrained by time. For example, consider the placement of MCL_1 in Figure 8.4 for the following four cases.

1. **Data arrives at the MCL_1 input at the *end* of L2 phase time: $t_{S1a} = 0$.**

   If S1a is moved in front of MCL_1, then the input data arrives at the end of the L2 phase time. This removes the "degree of softness" at the L2-L1 phase boundary and it becomes "hard." Timing becomes critical: clock jitter and skew may reduce the L2 phase time or circuit delay variations may increase the L2 phase delay, delaying the input signal to MCL_1 beyond the positive level of the C_2 clock. This will cause MCL_1 to latch the wrong information (or, similarly, increase the set-up requirements of the MCL output to the S1b logic). Therefore, for MCL_1 to latch proper information, the L2 phase time may need to be increased to overcome the clock jitter and skew time (t_{jitter} and t_{skew}) which increases the machine cycle time, degrading performance.[15]

2. **Data arrives at the MCL_1 input at the *beginning* of L2 phase time: $t_{S2b} \sim t_{gap12}$.**
   If the S2b logic is on the order of the time gap between the C_1 and C_2 positive levels, then the input data arrives at MCL_1 at the beginning of the L2 phase time. Timing again becomes critical: clock jitter and skew may cause the input to arrive

15. Or, equivalently, the actual frequency of the machine is less than the desired frequency (see equation (8.8)).

at MCL_1 before the C_2 clock. This will introduce a latch relaunch penalty and, potentially, a small amount of dead time. That is, the S2b logic must wait for the opening of MCL_1 before input data arrives to evaluate, which wastes valuable evaluation time and may also increase the machine cycle time.

3. **Data arrives at the MCL_1 input *before the beginning* of L2 phase time.**

If the data arrives at MCL_1 before the beginning of the L2 phase time, then a "latch relaunch penalty" is inevitable, which increases the cycle delay. That is, the input to MCL_1 will arrive before the C_2 clock, causing the S2b logic to wait for the opening of MCL_1 to evaluate, again wasting valuable evaluation time. Additionally, there is a potential cycle time violation: if MCL_1 is moved ahead of S2a, then a cycle time failure may occur because the cycle delay exceeds the maximum limit. That is, if the total cycle delay exceeds $1*t_{cycle}$, then the machine will not operate at the desired cycle time because slack borrowing has been prevented by the poor placement of MCL_1.

4. **Data arrives at the MCL_1 input *during* the L2 phase time.**

MCL_1 should be positioned so that the input signal arrives when the C_2 clock is high. This will remove any latch relaunch and clock skew penalties from the datapath, producing the optimum machine performance by reducing the cycle time. The amount of "softness" at both boundaries (S2b and S1a) should be sufficiently large to account for clock signal edge inaccuracy. That is, the logic both in front of and after MCL_1 should be of sufficient delay to remove relaunch and skew from the timing. Therefore, there exists a *minimum* delay requirement for the logic circuitry S2b and S1a and the latch MCL_1 to keep the input to MCL_1 during the L2 phase time. The delay, then, must exceed the sum of the gap times and the latch delay

$$\text{minimum}\{t_{S2b} + t_{S1a} + t_{L2}\} = t_{gap12} + t_{gap21} + t_{L2}. \tag{8.28}$$

By removing the latch delay from both sides, this reduces to

$$\text{minimum}\{t_{S2b} + t_{S1a}\} = t_{gap12} + t_{gap21} \tag{8.29}$$

which is applicable to *any* L2 phase. Additionally, equation (8.29) can be extended from the L2 phase for the minimum delay in any L1 phase time as well:

$$\text{minimum}\{t_{S1b} + t_{S2a}\} = t_{gap12} + t_{gap21}. \tag{8.30}$$

This means the clock edges in both phase times will always be *at least* coincident with the logic evaluation.

Equations (8.29) and (8.30) provide a formula for latch position within the phase time. However, if the minimum of these equation is met, any slight variation in logic evaluation time or clock arrival time will cause a relaunch penalty or a dead time to occur. Consequently, to add a small amount of design margin, a safer rule for the minimum logic delay in phase time is twice the sum of the gap times, or

$$\text{minimum logic delay in a given phase time} = 2*(t_{gap12} + t_{gap21}). \qquad \textbf{(8.31)}$$

This gives rise to a corresponding phase partitioning rule:

> **Phase Partitioning Rule:** To safely avoid relaunch penalties, the logic delay in a given phase time should be a minimum of twice the sum of the gap times.

The key to phase partitioning, then, is to position the latches within the particular phases such that the data arrival time is always during the corresponding phase time. This requires both maintaining a minimum amount of logic between latches (equation (8.31)) and obeying the phase slack borrowing rules.

> **Phase Partitioning Rule:** The optimum placement for latches is sufficiently inside the phase times to prevent both dead time and relaunch penalties.

However, once inside the optimum timing window, the final position for the latches should not be chosen randomly. Rather, their final position should be chosen based on key design considerations such as delay, area, and power.

For delay considerations, the latches should be positioned such that the overall cycle delay is minimized. In addition to avoiding relaunch penalties and dead times, a latch may be used for gain, e.g., replacing buffers, which decreases overall delay by reducing the latch delay impact.

Phase partitioning may reduce the latch count, drive requirements, wiring crossovers, dual-rail logic, etc., which reduces area. For example, moving a latch to the output side of a four-way mux from the input will reduce the overall latch count for the data signals from four to three, which is a 75% savings!

Power reduction can be achieved using phase partitioning techniques. That is, latch movement may reduce output drive and/or loading requirements, input loading, wiring, dual rail logic, etc. For example, moving a latch to provide a complement function to eliminate dual rail logic may reduce power by 50%!

Because of their complex nature, phase partitioning and slack borrowing are best learned by examples (see Section 8.2.4 on page 306). However, before moving to them, it is important to expand the discussion of slack borrowing and phase partitioning beyond the 50% symmetric duty cycle clocks.

8.2.3 Slack Borrowing and Phase Partitioning in Asymmetric 2-Phase Systems

Slack borrowing and phase partitioning in an asymmetric two-phase system are simple extensions to the symmetric 50% duty cycle case.

Asymmetric duty cycle two-phase systems are actually quite rare in industry, primarily because they divide the cycle into unbalanced segments. This can create confusion: the half-cycle which a logic circuit resides in forces different constraints on that circuit. Figure 8.6 shows a sample clock diagram for such a system.

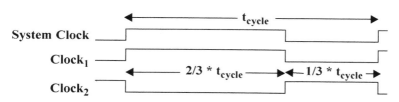

FIGURE 8.6 A Sample Asymmetric 2-Phase System Clock System

In the sample system of Figure 8.6, the system clock has been split up into two phases with separate duty cycles. In the L1 phase, the total available time is two-thirds of the cycle, while the L2 phase receives only one-third of the cycle time. Note that any division is possible, within reason.[16]

In terms of slack borrowing, the asymmetric system has several effects. First, the L1 and L2 phase times are no longer equal, even though the overall cycle time remains the same. Thus, some of the equations derived for the symmetric case remain intact, while others do not.

In terms of the maximum amount of time that can be borrowed for an L2 phase delay from an L1 phase time, the asymmetric case changes equation (8.17) to

$$\text{maximum}\{t_{S2a}\} = t_{pL1} \qquad (8.32)$$

because the L1 phase time is no longer one-half of a cycle time. This changes the maximum cycle delay (see equation (8.18)) to

$$\text{maximum}\{t_{dcycle}\} \cong t_{cycle} + t_{pL1}. \qquad (8.33)$$

---

16. A clock phase could not have a duty cycle smaller than the uncertainty in the clock edges.

Note that this may be greater or less than that of equation (8.18) depending on the L1 phase time. Furthermore, equation (8.32) means the maximum L2 phase delay is

$$\text{maximum}\{t_{dL2phase}\} \cong t_{cycle} + t_{pL1} \qquad (8.34)$$

Note that this amount may also be greater or less than that of the symmetric case. Additionally, note that the L1 phase delay is given by an extension to equation (8.22) and is now

$$\text{maximum}\{t_{dL1phase}\} \cong t_{cycle} + t_{pL2}. \qquad (8.35)$$

If we consider the clocking system example of Figure 8.6, these equations mean that

$$\text{maximum}\{t_{dcycle}\} \cong 1.67*t_{cycle} \qquad (8.36)$$

$$\text{maximum}\{t_{dL2phase}\} \cong 1.67*t_{cycle} \qquad (8.37)$$

$$\text{maximum}\{t_{dL1phase}\} \cong 1.33*t_{cycle}. \qquad (8.38)$$

Additionally, note that the overall delay of an n-stage pipeline must still obey equation (8.19). That is, the maximum time used the machine still cannot exceed to total available, regardless of the clock phase time assignments.

As for phase partitioning, the asymmetric system has no effect on the minimum delay inside a phase time of equation (8.31) as the gap times are not affected. However, the amount of time that a latch may be moved within a phase changes. Remember that phase partitioning allows a latch to be moved around inside a phase time. Since the phase time is either shorter or longer in the asymmetric system, the amount of possible latch movement is likewise.

The net effect of the asymmetric system on slack borrowing and phase partitioning, then, is relatively small. That is, the overall delay of the machine pipeline cannot change, or equation (8.19) must still hold, even though the amount of time borrowed across phases and cycles varies, as does the maximum movement of latches.

8.2.4 Slack Borrowing Examples

Slack borrowing examples are the best way to understand the concepts.

Because of the complex and somewhat counter-intuitive nature of the subject, examples are the best way to understand the concepts of slack borrowing and phase partitioning. This section presents three such examples using a symmetric 50% duty cycle two-phase system.[17]

The first example is shown in Figure 8.7 on page 307. Here $cycle_2$ is borrowing one-quarter cycle from $cycle_1$, which is borrowing one-quarter cycle for itself *plus* one-quarter cycle for $cycle_2$ for a *total* of one-half cycle from $cycle_0$. The cycle delays are

$$t_{dcycle-0} = 0.5*t_{cycle} \qquad\qquad (8.39)$$

$$t_{dcycle-1} = 1.25*t_{cycle} \qquad\qquad (8.40)$$

$$t_{dcycle-2} = 1.25*t_{cycle} \qquad\qquad (8.41)$$

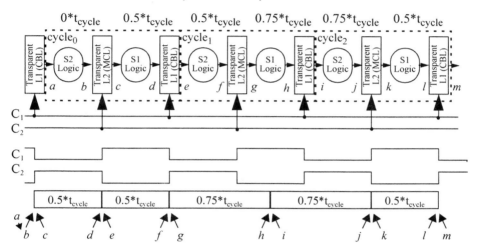

FIGURE 8.7 Slack Borrowing Example 1

which are consistent with equation (8.18). The total delay for the three cycles is

$$t_{total-delay} = 3*t_{cycle} \qquad\qquad (8.42)$$

which is consistent with equation (8.19).

However, suppose that point *l* is 10 [ps] late. Because the CBL between *l* and *m* receives its inputs at the end of the L1 phase time, the L1 phase time must increase to latch the proper state. A cursory analysis would say this time is 10 [ps], which increases the cycle time by 10 [ps]. However, if the system clock is slowed, all subsequent clock edges will be delayed (move to the right).[18] Any latch at the beginning of

---

17. In these examples we also consider the latch delay as part of the phase delay.
18. See the 2-phase clocking diagram of Figure 8.1 on page 288.

a phase time will suffer a relaunch penalty, such as the MCL for cycle_1. Additionally, any latch at the end or inside of a phase time will not suffer a relaunch penalty, such as the MCL for cycle_2. Consequently, the delay across cycles 1 and 2 can be *averaged* and the total impact on the machine cycle time is (10 [ps])/(2) or 5 [ps]. This means the overall machine cycle time is

$$t_{cycle} = t_{cycle\text{-}desired} + 5 \; [ps]. \tag{8.43}$$

Figure 8.8 on page 308 shows a second example, consisting of a timing diagram applicable to the system of Figure 8.7. Here cycle_0 borrows one-quarter cycle from the previous cycle. Because it only requires one cycle time, cycle_0 passes this time to cycle_1, which also only requires one cycle time, and, thus, passes the time to cycle_2. This cycle uses the borrowed time because it requires one and one-quarter cycle. The

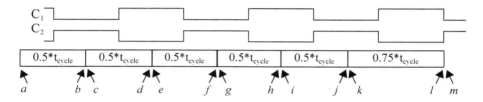

FIGURE 8.8 Slack Borrowing Example 2

cycle delays in this example are

$$t_{dcycle\text{-}0} = 1 * t_{cycle} \tag{8.44}$$

$$t_{dcycle\text{-}1} = 1 * t_{cycle} \tag{8.45}$$

$$t_{dcycle\text{-}2} = 1.25 * t_{cycle} \tag{8.46}$$

which are consistent with equation (8.18). The total delay for the three cycles is

$$t_{total\text{-}delay} = 2.25 * t_{cycle} \tag{8.47}$$

which is consistent with equation (8.19).

Note, however, that in this example all latches are located in the middle of their associated phase times, creating "soft" latch boundaries. Assume, again, that l to m is 10 [ps] late. Because the latch boundaries are soft, moving the clock edges will create no latch relaunch penalties. Consequently, the total delay impact on cycle time can be averaged to (10 [ps])/(3), or 3.33 [ps]. This means the resulting cycle time is

$$t_{cycle} = t_{cycle\text{-}desired} + 3.33 \text{ [ps]}. \tag{8.48}$$

These last two examples show that if additional time is required in a given cycle(s) and no more slack borrowing is available, then the amount of additional time required can be averaged across all sequential cycle that will avoid a relaunch penalty. This, then, means that the *minimum* achievable cycle time in an *n*-stage pipeline that requires *x* additional time is

$$t_{cycle} = t_{cycle\text{-}desired} + x / n. \tag{8.49}$$

Or, equivalently, we can rewrite this as

$$t_{cycle} = (n * t_{cycle\text{-}desired} + x) / n. \tag{8.50}$$

This gives rise to a slack borrowing rule concerning time averaging:

Slack Borrowing Rule: If additional time is required in a pipeline or sub-pipeline, slack borrowing permits the averaging of the additional time by moving clocks until such time that a latch relaunch penalty is required.

Furthermore, this provides a rule on the minimum achievable cycle time:

Slack Borrowing Rule: The minimum achievable cycle time in a pipeline that requires additional time is the sum of the desired cycle time and the additional time divided by the pipeline length, or, equivalently, is the dividend of the number of stages times the desired cycle time plus the additional time and the number of stages.

Figure 8.9 shows a third example where the $cycle_0$ result (*e*) is available at three-quarters into the cycle time and the $cycle_1$ result (*i*) use an additional three-quarters. Subsequently, the S2 logic of $cycle_2$ evaluates in a quarter-cycle. But, note that the MCL of $cycle_2$ is not open until another quarter cycle later. This creates a dead time in the total path delay (equal to that same quarter-cycle) for $cycle_2$.

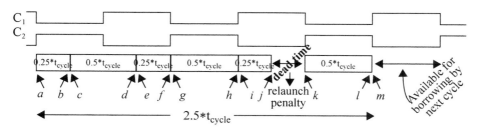

FIGURE 8.9 Slack Borrowing Example 3

The cycle delays are

$$t_{dcycle-0} = 0.75*t_{cycle} \qquad (8.51)$$

$$t_{dcycle-1} = 0.75*t_{cycle} \qquad (8.52)$$

$$t_{dcycle-2} = 1*t_{cycle}. \qquad (8.53)$$

The total delay for the three cycles is

$$t_{total-delay} = 2.5*t_{cycle} \qquad (8.54)$$

which includes the additional penalty for the latch relaunch. To avoid the relaunch penalty, phase partitioning techniques could be applied to reapportion the S2 and S1 logic of cycle$_2$ such that

$$t_{L1phase} \geq 0.5*t_{cycle} \qquad (8.55)$$

$$t_{L2phase} = 0.75 - t_{L1phase}. \qquad (8.56)$$

This would reduce the delay of cycle$_2$ to

$$t_{dcycle-2} = 0.75*t_{cycle} \qquad (8.57)$$

and the total path delay would become

$$t_{total-delay} = 2.25*t_{cycle}. \qquad (8.58)$$

Note that equation (8.58) is well below the limit of the three times the cycle time as required by equation (8.19). Therefore, the application phase partitioning techniques has reduced the total delay by one-quarter of a cycle!

8.2.5 Looping Effects on Slack Borrowing

Loops (feedback) inside logical pipelines force the application of slack borrowing rules to each individual loop.

Real chip design environments inevitably require internal looping of paths. That is, a state machine inside a design, such as a microprocessor, nearly always requires input from a previous operation/state to determine the next. This input may come from the next stage in the pipeline or many down the pipeline. Called "loops" because they create return paths in the logical flow, they complicate the analysis of slack borrowing by creating additional requirements on the system. That is, each individual loop must now be considered a separate machine in itself. For example, consider the example of Figure 8.10 on page 311 where the output of cycle$_2$ feeds the S2 logic of cycle$_1$.

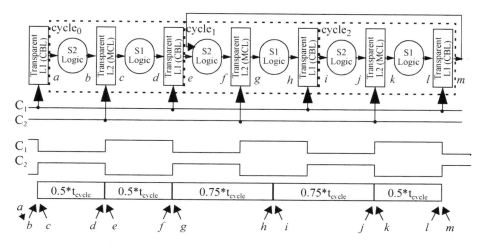

FIGURE 8.10 Slack Borrowing Example #4 (Including Looping Considerations)

In the previous three examples without feedback, the slack borrowing rules required that the sum of the cycle delays of cycles 0, 1, and 2 be less than or equal to three times the cycle time:

$$t_{dcycle0} + t_{dcycle1} + t_{dcycle2} \leq 3 * t_{cycle} \qquad (8.59)$$

This must still hold for Figure 8.10. However, in this case the loop from $cycle_2$ back to $cycle_1$ forces the slack borrowing rules to apply across these last two cycles. Otherwise, the subsystem of cycles 1 and 2 will not operate in the desired cycle time. Therefore, the system of Figure 8.10 also requires

$$t_{dcycle1} + t_{dcycle2} \leq 2 * t_{cycle} \qquad (8.60)$$

Applying the delays of Figure 8.10 forces two cycle time requirements:

$$t_{dcycle0} + t_{dcycle1} + t_{dcycle2} = 0.5 * t_{cycle} + 1.25 * t_{cycle} + 1.25 * t_{cycle} = 3 * t_{cycle} \qquad (8.61)$$

$$t_{dcycle1} + t_{dcycle2} = 1.25 * t_{cycle} + 1.25 * t_{cycle} = 2.5 * t_{cycle} \qquad (8.62)$$

However, note that the sum of the cycle delays in equation (8.62) exceeds the allowable time. This means that the second equation will limit the actual cycle time to a *minimum* of the total cycle delay divided by the number of stages in the loop, or using equation (8.50)

$$\text{minimum } t_{cycle} = (time)/(\# \text{ stages}) = 2.5 * t_{cycle\text{-}desired} / 2 = 1.25 * t_{cycle\text{-}desired}. \quad \textbf{(8.63)}$$

This means that the actual machine frequency is reduced, or

$$f_{actual} \leq 1/t_{cycle\text{-}actual} = 0.8 * f_{desired}. \quad \textbf{(8.64)}$$

Thus, the machine frequency in the example has reduced a *minimum* of 20% because of the loop between cycles 1 and 2!

In summary, each loop in a system introduces an additional constraint on slack borrowing potentially limiting machine performance. That is, every loop must comply with the slack borrowing rules or else, per equation (8.50), the overall machine cycle time will be increased. This gives rise to the final rule for slack borrowing.

> **Slack Borrowing Rule:** Internal loops must follow the slack borrowing rules as if they were a machine (pipeline) in themselves.

8.3 Time Stealing

Time Stealing gains evaluation time by taking it from the *next* cycle and is ideally suited for dynamic logic in two-phase system or static logic in a master-slave system.

Time stealing refers to the cases where a logical partition requires additional time for evaluation but cannot borrow left over time (if any) by the previous cycle or phase. As a result, the logical partition is forced to steal the time from the subsequent cycle/phase, which leaves that cycle/phase with less evaluation time.[19] The stealing of additional time is considered involuntarily surrendered and is, generally speaking, obtained by adjusting clock arrival times.

Because additional evaluation time is obtained by adjusting clock arrival times, time stealing opportunities arise in applications using edge-triggered logic, such as dynamic, and/or latches, such as master-slave. That is, if a clock edge is moved beyond the standard arrival time, then the edge-triggered circuitry will necessarily gain additional evaluation time. Note that because the circuits are edge-triggered, slack may not be borrowed as the clock must arrive before any evaluation can occur.

Because two-phase clocking is ideal for high-speed designs,[20] time stealing is ideally suited for dynamic logic[21] in such a system using separated latch techniques. Addi-

---

19. Unless, of course, that cycle steals from its subsequent cycle.
20. See Section 7.6.

tionally, in a master-slave style design utilizing static logic,[22] time stealing techniques can be used to gain additional evaluation time.

> **Rule of thumb:** Time stealing is ideally suited for dynamic logic in a two-phase clocking system utilizing separated latch design techniques, or static logic in a master-slave system.

An additional extension to time stealing can be made to include using time left over in a previous cycle by adjusting the clock to arrive earlier. That is, if a path utilizes edge-triggered techniques (logic or latches) and the previous cycle has left over time, then using an earlier arriving clock will allow the use of that time. Note that because this time is from the previous cycle, it could be confused with slack borrowing. However, the use of an earlier clock means a clock edge arrival time adjustment, which by definition violates a basic premise of slack borrowing (see page 286).

Historically, time stealing has been off-handedly called "cycle stealing." However, this term is somewhat of a misnomer. That is, just as with slack borrowing, time stealing in a two-phase system has can be divided into two separate cases: *cycle time stealing* and *phase time stealing*.

Cycle time stealing permits edge-triggered pipelines between cycle boundaries to use more than one cycle time and still fit within a single cycle clock boundary *while* maintaining the overall machine cycle time. That is, if the time used for logic evaluation in an edge-triggered cycle boundary *exceeds* one cycle and the machine still works at speed, then evaluation time has been stolen from the subsequent cycle.

Similarly, *phase time stealing* allows edge-triggered pipeline partitions in a particular clock phase (typically one-half of a cycle) to use more time than is readily available and *still* maintain the overall machine cycle time. That is, if the time used for logic evaluation in a particular clock phase of an edge-triggered system *exceeds* the clock phase time and the machine still works at speed, then phase time has been stolen from the subsequent phase.

To fully understand cycle and phase time stealing, we first consider dynamic logic in a symmetric 50% duty cycle two-phase system with non-overlapping clocks, then extend the ideas into an overlapping symmetric two-phase system and an asymmetric two-phase system. This is followed by a section on cycle stealing for master-slave systems. Finally, time stealing from the previous cycle is discussed. Note that examples of time stealing are provided inside each section (where necessary) instead of

---

21. See, for example, Section 3.2.
22. See, for example, Section 2.2.

---

inside a separate section (as in the slack borrowing discussion) because the concepts of time stealing are not as difficult as slack borrowing.

It should be pointed out that although the discussions on time stealing below utilize dynamic logic, identical techniques could be applied to systems using static logic as in the slack borrowing discussions.

8.3.1 Time Stealing for Dynamic Logic in Symmetric 50% Duty Cycle 2-Phase Non-Overlapping Systems

In a two-phase non-overlapping system, time stealing for dynamic logic is limited by the amount of additional evaluation time applied to one or both of the clock phases.

As discussed in Section 7.6.1, a symmetric 50% duty cycle two-phase system is one in which the centrally generated and distributed system clock is split into two identical but logically inverted clocks. Figure 8.1 on page 288 showed an idealized clock timing diagram.

Additionally, a non-overlapping system is one in which the clock signals are not simultaneously active, also as shown in Figure 8.1.

As in slack borrowing, the clock signal denoted "C_1" in Figure 8.1 is typically defined as the "cycle boundary clock," while "C_2" is the "mid-cycle boundary clock." As before, the definitions are essentially arbitrary, but differentiate between cycle time stealing and phase time stealing. That is, swapping the definitions is effectively meaningless.[23]

In a chip application, the clocks of Figure 8.1 on page 288 are used to control a two-phase system as shown in Figure 8.11.

Note that the clocks control transparent latch elements alternately, separated by dynamic logic circuitry (noted "Dyn. Logic"). This creates a pipeline structure.[8.5] Furthermore, note that the logic and the latch it feeds utilize the same clock.This creates a system where the latches are data-gated (open) during the logical evaluation time, as opposed to an edge-triggered latch situation, and simplifies the design.

Because the logic circuits are edge-triggered, each dynamic circuit cannot begin its evaluation before its associated clock signal is active, regardless of the input signal stability. Therefore, any slack time in a phase is automatically "dead time."

---

23. Unless special extensions are made to the clock definitions for testability purposes.[8.4]

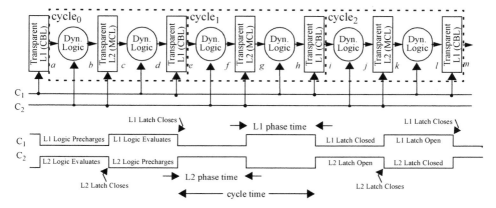

FIGURE 8.11 2-Phase System using Dynamic Logic and Non-Overlapping Symmetric 50% Duty Cycle Clocks

Time Stealing Rule: Dead time penalties are unavoidable in an edge-triggered design, such as dynamic logic in a 2-phase system.

Additionally, because it requires an active clock, evaluation cannot continue past the falling (inactive) edge of the clock. As a result, any dynamic logic circuit receives a maximum evaluation time equal to the time its clock is active, which is equal to the phase time in which it resides.

Thus, for the system of Figure 8.11 (with no clock arrival time adjustments) the maximum *phase delay* is equal to the *phase time*, or in the terms defined in Section 8.2.1,

$$\text{maximum}\{t_{dL2phase}\} = t_{pL2} \tag{8.65}$$

$$\text{maximum}\{t_{dL1phase}\} = t_{pL1} \tag{8.66}$$

Because each phase time in a standard symmetric 50% duty cycle system is one-half of a cycle time,[24] these can be rewritten as

$$\text{maximum}\{t_{dLxphase}\} = 0.5*t_{cycle}. \tag{8.67}$$

Per equation (8.11), the cycle delay is equal to the sum of the phase delays, or

$$\text{maximum}\{t_{dcycle}\} = \text{maximum}\{t_{dL2phase}\} + \text{maximum}\{t_{dL1phase}\} = 1*t_{cycle}. \tag{8.68}$$

---

24. See equations (8.2) and (8.4).

Slack Borrowing and Time Stealing **315**

Additionally, because of the edge-triggered nature of the dynamic circuits if an input arrives before the clock is active the evaluation cannot begin. Consequently, the minimum phase delays of the system of Figure 8.11 (with no clock adjustments) are

$$\text{minimum}\{t_{dL2phase}\} = t_{pL2} \qquad \text{(8.69)}$$

$$\text{minimum}\{t_{dL1phase}\} = t_{pL1}. \qquad \text{(8.70)}$$

<u>**Time Stealing Rule:**</u> If clock edges are not adjusted, then both the minimum and maximum phase delays are equal to the phase time.

This means that the minimum cycle delay is

$$\text{minimum}\{t_{dcycle}\} = 1*t_{cycle}. \qquad \text{(8.71)}$$

Equations (8.68) and (8.71) mean that the minimum and maximum *cycle delays* are equal to the *cycle time*, assuming the clock arrival times are not adjusted.

<u>**Time Stealing Rule:**</u> If clock edges are not adjusted, then both the minimum and maximum cycle delays are equal to the cycle time.

If we compare these results for time stealing to those for slack borrowing, it is evident that there can be no *automatic* use of left-over time. Remembering that each circuit requires an active clock signal to evaluate, this means that if a dynamic circuit requires additional time to evaluate, it must increase its phase time by widening the active clock time. This can only be done by moving either the active (rising) edge earlier, or the inactive (falling) edge later.

<u>**Time Stealing Rule:**</u> To perform time stealing, the phase time must be increased by widening the active clock time.

Assuming the clock signals are driven from a centrally routed clock distribution system, the clock edge arrival times are set at the circuit and latch boundary. This prevents moving the active (rising) edge of the phase clock earlier in time to widen a phase time.[25] However, it does not mean that the inactive (falling) edge cannot be moved later in time. Thus, any additional evaluation time can only be obtained by delaying the inactive edge of the phase clock, which, then, increases the phase time.

<u>**Time Stealing Rule:**</u> A phase time increase is achieved by delaying the inactive clock edge.

---

25. It is possible in *some* clock distribution systems to use an earlier clock to obtain an earlier rising (active) clock edge to widen the active clock time. This technique is described in Section 8.3.5, but it not *generally* performed.

We can further extend this concept to a cycle by noting that the cycle delay is the L2 phase delay plus the L1 phase delay.[26] Thus, if a cycle delay is to exceed the limit set by equation (8.71), then either the last phase time (L1) or both phase times (L2 and L1) must increase. Note that if the L2 phase time alone increases, then time is stolen from the L1 phase time from the same cycle and the cycle delay does *not* increase.

Time Stealing Rule: A cycle delay increase is achieved by moving (delaying) the inactive clock edge of either the L2 phase alone or both the L2 and L1 phases.

Consider, though, what effect of increasing a phase time has on the subsequent phase. If a phase time has been increased, then in the *next* phase the dynamic logic circuitry cannot begin its evaluation until both its clock is active *and* its inputs are stable. To clarify this, consider the example of Figure 8.12 on page 317.[27]

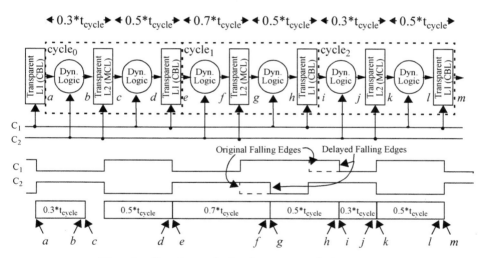

FIGURE 8.12 Time Stealing Example for the System of Figure 8.11

Note that the L2 phase of $cycle_0$ does not use all of the available phase time. However, because the logic in the subsequent L1 phase time is edge-triggered, this time cannot be used. Thus, since no clock edges are moved for this phase, its delay must follow equations (8.67) and (8.69):

---

26. Using equation (8.11).

27. Note that the latch delay times have been included in the logic delay times to simplify the figure, which was also done for the slack borrowing examples.

$$t_{dL2phase-cycle0} = 0.5*t_{cycle}. \tag{8.72}$$

The subsequent L1 phase of $cycle_0$ also uses a full normal phase time, so its delay is

$$t_{dL1phase-cycle0} = 0.5*t_{cycle}. \tag{8.73}$$

However, the L2 phase of $cycle_1$ requires 20% more evaluation time than a normal phase. Consequently, the falling edge of the C_2 clock is delayed by the additional time required, stealing time from the next L1 phase. This creates an extended phase time equal to the regular phase time plus the additional time. Using equation (8.65), this means the L2 phase delay is

$$t_{dL2phase-cycle1} = t_{pL2} = 0.5*t_{cycle} + 0.2*t_{cycle} = 0.7*t_{cycle}. \tag{8.74}$$

The subsequent L1 phase of $cycle_1$ requires a full (normal) phase time. However, the use of time stealing in the L2 phase has shortened the available time by the additional amount it used. Thus, the falling edge of the C_1 clock of the L1 phase in $cycle_1$ must be delayed by the same amount as the C_2 clock, which means that the L1 phase steals time from the next L2 phase. This creates a phase delay of

$$t_{dL1phase-cycle1} = 0.5*t_{cycle}. \tag{8.75}$$

Furthermore, the L2 phase of the next cycle ($cycle_2$) requires only 30% of the normal phase time, which is equivalent to the normal phase time less the additional amount required for the L2 phase of $cycle_1$. This means its phase delay is

$$t_{dL2phase-cycle2} = 0.3*t_{cycle}. \tag{8.76}$$

The last phase delay in the example of Figure 8.12 requires a normal phase time, or

$$t_{dL1phase-cycle2} = 0.5*t_{cycle}. \tag{8.77}$$

Thus, the cycle delays are

$$t_{dcycle0} = 1.0*t_{cycle} \tag{8.78}$$

$$t_{dcycle1} = 1.2*t_{cycle} \tag{8.79}$$

$$t_{dcycle2} = 0.8*t_{cycle}. \tag{8.80}$$

This produces a total delay for the three cycle pipeline of three times the cycle time.

Looking at the results, the L2 phase of $cycle_1$ increases its evaluation time by phase stealing from the subsequent L1 phase (of $cycle_2$). The L1 phase of $cycle_2$, in turn, uses phase stealing from the subsequent L2 phase (of $cycle_3$) to obtain more evaluation time.

Time Stealing Rule: A phase may only steal time from the next phase in the machine pipeline.

The delay of $cycle_1$ is now greater than the cycle time, which means cycle time stealing has occurred, and that time has been taken from $cycle_2$.

Time Stealing Rule: A cycle may only steal time from the next cycle in the machine pipeline.

Furthermore, note that the additional time obtained for $cycle_1$ meant its L1 phase had to phase steal from the L2 phase of $cycle_2$. In the example, the L2 phase of $cycle_1$ also required more time, which it stole from its own L1 phase. Note that the latter is not required for cycle time stealing. That is, if the L2 phase only required a normal phase time and the L1 phase required additional time, then stealing could also be performed.

Time Stealing Rule: Cycle time stealing requires that phase time stealing has occurred in its L1 phase or both its L1 and L2 phases.

As in slack borrowing, there are limits on the amount of time that can be stolen by a phase or a cycle. However, the limits are not nearly as constrained. That is, the clock signals to the dynamic circuits and latches can be continually delayed for each subsequent phase such that a *nearly* any amount time can be used for a given cycle. Note that this requires each rising and falling edge to be delayed to every latch and dynamic circuit.

To explore the limits on time stealing, note that the maximum delay of n times the cycle time may be used in an n-stage pipeline,[28] or

$$\text{minimum delay across } n \text{ cycles} = n * t_{cycle}. \tag{8.81}$$

Time Stealing Rule: The sum of all cycle delays across an n-stage pipeline machine can be a maximum of n times the cycle time.

Furthermore, for the pipeline to exist note that all n cycle latches (CBLs and MCLs) must exist.[29] If clock movement is allowed, then equations (8.69) and (8.70), which describe the minimum phase delays, no longer apply. Rather, each phase must consist of at least a latch (CBL or MCL). This means that the minimum phase delay in a clock *adjusted* system is

$$\text{minimum}\{t_{dLxphase}\} = t_{Lx}. \tag{8.82}$$

---

28. Note that this was also a requirement for slack borrowing (see equation (8.19) on page 297).

29. Else, it would not be an n-stage pipeline.

Slack Borrowing and Time Stealing **319**

Time Stealing Rule: The minimum phase delay in a clock adjusted system is equal to the phase latch delay.

Since a cycle delay is the sum of the phase delays, its minimum is

$$\text{minimum}\{t_{dcycle}\} = t_{L1} + t_{L2}. \tag{8.83}$$

Time Stealing Rule: The minimum cycle delay in a clock adjusted system is equal to the sum of the phase latch delays.

Using equation (8.81) this means that across n cycles, the total minimum delay is

$$\text{minimum delay across } n \text{ cycles} = n * (t_{L1} + t_{L2}). \tag{8.84}$$

Subtracting equation (8.84) from (8.81) gives the maximum delay that can used by the logic circuitry in an n stage pipeline:

$$\text{maximum } \textit{logic circuity delay} \text{ across } n \text{ cycles} = n * (t_{cycle} - t_{L1} - t_{L2}). \tag{8.85}$$

Time Stealing Rule: The maximum logic circuitry delay in a clock adjusted n-stage pipeline machine is n times the quantity of the cycle time minus the phase latch delays.

If this is used by a single cycle in the pipeline, then the maximum cycle delay is the maximum logic circuitry delay (given by equation (8.84)) plus the delay of the latch circuitry in a single cycle (given by equation (8.83)) or

$$\text{maximum}\{t_{dcycle}\} = n * (t_{cycle} - t_{L1} - t_{L2}) + t_{L1} + t_{L2}. \tag{8.86}$$

Re-writing,

$$\text{maximum}\{t_{dcycle}\} = n * t_{cycle} - (n-1)*(t_{L1} + t_{L2}). \tag{8.87}$$

Time Stealing Rule: The maximum cycle delay in a clock adjusted n-stage pipeline machine is n times the cycle time minus n -1 times the sum of the phase latch delays.

Furthermore, if the maximum logic circuitry delay is used by a single phase inside that cycle, then the maximum phase delay can be found by adding the phase latch delay to equation (8.85):

$$\text{maximum}\{t_{dLxphase}\} = n * (t_{cycle} - t_{L1} - t_{L2}) + t_{Lx}. \tag{8.88}$$

If this is in an L1 phase, then the maximum phase delay is

$$\text{maximum}\{t_{L1phase}\} = n * (t_{cycle} - t_{L2}) - (n-1)*t_{L1}. \tag{8.89}$$

Similarly, if this is in an L2 phase, then the maximum phase delay is

$$\text{maximum}\{t_{L2phase}\} = n * (t_{cycle} - t_{L1}) - (n-1)*t_{L2}. \tag{8.90}$$

Time Stealing Rule: The maximum phase delay in a clock adjusted n-stage pipeline machine is n times the quantity of the cycle time minus the opposite phase latch delay, minus n -1 times the phase latch delay.

It should be noted that using all the available time in a single cycle or phase is not as simple to perform as equations (8.86) through (8.90) may appear. That is, delaying the clock is inherently inaccurate because of tolerances across process and environment.[30] Additionally, the accuracy of the clock edge is at best within 5-10%.[31] If these effects are accounted for in the system of Figure 8.11 and clock edge adjustment techniques are applied to allow a *single* cycle or phase to use *all* the available evaluation time, then there must exist a minimum delay between latches to prevent a flush situation from occurring. That is, if in the system of Figure 8.11 the latches are moved adjacent to one another, then the clock edge inaccuracy may permit multiple latches to be open simultaneously. Because these are adjacent, the pipeline will lose state.

Time Stealing Rule: In a clock adjusted system, if all the available evaluation time is consumed by a single phase or cycle, then the designer must be careful to prevent a flush problem from occurring because of the adjacent latches and the clock skew, adding in the additional uncertainty introduced by the delay circuitry used.

Additionally, if such a system is created, it is important to note that each latch requires a minimum clock pulse width to store information.[32] This has the effect that the minimum phase delay of equation (8.82) and cycle delay of equation (8.83) may not be realistically achievable. That is, if the minimum clock pulse width for the latches is larger than the data-gated delay, then minimum phase delay must be increased to the minimum clock pulse width. This affects the minimum cycle delay as it is the sum of the minimum phase delays.

Time Stealing Rule: In a clock adjusted system, if the minimum clock pulse width for the latches to reliably store information is larger than the data-gated delay, then the actual minimum phase delay is equal to that pulse width.

A further note should be made about widening the active time of a clock to gain more evaluation time for dynamic circuits relating to set-up times. If the extended evaluation time extends beyond the active time of the next phase clock, then the circuits in the two affected phases both evaluate during the time of the overlapped. Therefore, unless proper precharge levels are maintained at the input to the second phase, the problem of a premature evaluation can occur.[33]

---

30. See Chapter 1.
31. See Section 7.1
32. See Section 5.2
33. See, for example, Section 3.2.1.

Time Stealing Rule: In a clock adjusted system, if two successive phase clocks are simultaneously active due to the clock delay, then a potential premature evaluation problem can occur at the second phase.

Remedying this situation can be somewhat simple by delaying the rising active edge of the next phase an equal amount of time as the falling edge of the previous phase. For example, in the system of Figure 8.12 on page 317, if the active edge of the L1phase clock (C_1) of cycle$_1$ is delayed such that it is coincident with the falling edge of the C_2 clock of the previous phase, then the evaluation of the L1 phase dynamic circuitry cannot start until the previous L2 phase has completed. A similar technique can be applied for the L2 phase of the cycle$_2$. The resulting clock signal diagram is shown in Figure 8.13.

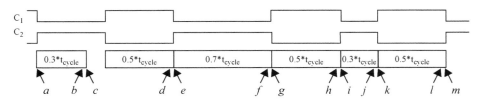

FIGURE 8.13 Clock Diagram for Figure 8.11 to Prevent Premature Evaluation

Note that this technique may or may be required based on the logic circuitry in an affected phase where time has been stolen.

Time Stealing Rule: In a clock adjusted system, if two successive phase clocks are simultaneously active due to the clock delay and a premature evaluation problem can occurs at the second phase, then the problem may be removed by delaying the active edge of the second phase an amount equal to the time stolen by the first phase.

The active time of the clocks must be sufficient to permit both the logical evaluation and the proper storage of information in the latch. However, note that delaying the active edge of the clock may result in a shorter precharge time. For example, in Figure 8.13, the precharge time of the L1 phase of cycle$_1$ has been shortened. If this amount of time is insufficient for the associated dynamic logic to precharge, then an incorrect evaluation will occur.[34] Thus, any delayed clock signal must also provide adequate precharge time.

Time Stealing Rule: In a clock adjusted system, any resulting internal clock must provide sufficient precharge time as well as evaluation time.

---

34. See, for example, Section 3.2.1 on page 93.

Furthermore, consider what would happen to the system of Figure 8.12 if point f were an additional 10 [ps] late. For the subsequent MCL to store the correct data, this means that the C_2 clock would have to be 10 [ps] late as well. Thus, if this problem were not discovered at design time (so as to use stealing techniques), then the additional time could be obtained from slowing the centrally generated clock signal. However, if both the C_1 and C_2 clocks are obtained from a centrally generated clock, slowing the C_2 by 10 [ps] will also slow C_1 by 10 [ps]. This, then, pushes out the total additional time to the system to 20 [ps], which means that the delay for cycle$_2$ is

$$t_{dcycle2} = t_{cycle} + 20 \text{ [ps]}. \tag{8.91}$$

Time Stealing Rule: In a two-phase edge-triggered system where the clocks are generated from a centrally driven signal, a system-wide delay in one clock phase causes an identical effect in the other. Thus, any system-wide delay in one edge results in a twice-paid penalty.

Note, though, that because the logic for the subsequent stages is edge-triggered, it cannot use any left-over (slack) time. That is, it must await the clock before evaluation can being. Therefore, if the global clock is delayed by 20 [ps], then the overall machine cycle time is also increased by the same 20 [ps], or

$$t_{cycle\text{-}actual} = t_{cycle\text{-}desired} + 20 \text{ [ps]} \tag{8.92}$$

which means that the machine frequency is slowed per equation (8.1) on page 291.

Comparing this to the slack borrowing discussion, there is no concept of time averaging for time stealing. That is, any additional time required for an n-stage pipeline cannot be averaged across any number stages.

Time Stealing Rule: In an edge-triggered system, time averaging is not possible.

Up to this point, we have ignored the issue of gap (separation) times between the clock signals.[35] That is, due to clock generation and routing, the clock signal edge arrival times have in inherent inaccuracy that affects time stealing. If the clock edges separate, then the time within the cycle that a phase begins evaluation is delayed by that separation (or gap) time. If this occurs, then this time automatically results in dead time because the logic is edge-triggered (see the discussion and rule on page 315).

However, it is possible to build into the clock system an overlap in the clock signals to prevent the dead time from occurring. That is, if the clocks are overlapped by an amount equal to the worst-case clock edge inaccuracy, then a dead time will never occur between phases.

---

35. See page 292 for more information on gap times.

Additionally, if across an entire chip design the worst cycle or phase requires a small amount of additional time beyond the standard time, it is possible to build-in an additional overlap into the clock signals beyond that to prevent dead times. That is, if the maximum amount of additional time required by all the cycles and phases is known and is sufficiently small, then the clock overlap tim can be designed in such a manner to normally overlap the clocks by that amount in addition to the clock uncertainty.

8.3.2 Time Stealing for Dynamic Logic in Symmetric 2-Phase Overlapping Systems

In an symmetric two-phase overlapping system, additional evaluation for dynamic logic is built into the clocking system and still permits time stealing.

A symmetric two-phase overlapping system is one in which the clocks are overlapped for a small amount of time. A sample system diagram is shown in Figure 8.14.

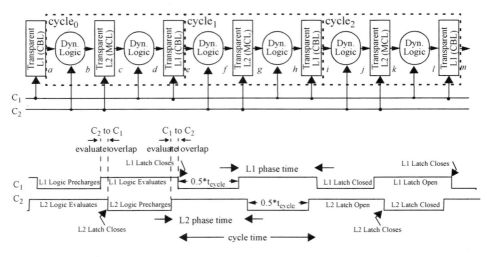

FIGURE 8.14 2-Phase System using Dynamic Logic and Overlapping Clocks

In this two-phase clocking style, the active phases of the clocks are overlapped by a small amount of time. This affects the overall machine and time stealing.[36]

---

36. See Section 7.6.1.

As discussed in Section 7.6.1, the possibility of a flushing problem in a machine using overlapped clocks is amplified. That is, because the different phase latches are simultaneously open for a small amount of time, any logically adjacent latches will cause a flush-through problem, resulting in the loss of the machine state. Consequently, in such a system the latches must be separated by a minimum amount of logic circuitry.

An overlapped system can have a major effect on time stealing. That is, for the example of Figure 8.14, note that the phase times are increased by the evaluate overlap times. Furthermore, note that each clock is normally *inactive* for one-half of a cycle time.[37] If the time of the C_2 to C_1 are simultaneously active is denoted "t_{C2C1}" and, similarly, "t_{C1C2}" for the C_1 to C_2 overlap, then in the sample system the phase times are increased beyond the one-half of a cycle time to

$$t_{pLx} = 0.5*t_{cycle} + t_{C2C1} + t_{C1C2} \qquad (8.93)$$

Since this is an edge-triggered system, then maximum phase delay for either phase is

$$\text{maximum}\{t_{dLxphase}\} = 0.5*t_{cycle} + t_{C2C1} + t_{C1C2}. \qquad (8.94)$$

Therefore, each phase actually has an *increased* evaluation time. If the amount of additional time required for a non-overlapped system is within this additional increase, then using an overlapped system will prevent the necessity of time stealing.

A cursory analysis would suggest that since the phase delays have increased, then the cycle delay has also increased as the cycle delay is the sum of the phase delays.[38] However, this is not true as the additional evaluation time actually occurs in an adjacent phase during the overlapping of the active clocks. Therefore, because any n-stage machine can use a maximum of n times the clock cycle time, using this time simply means that the next cycle(s) has less time available.

Additionally, the problem of dynamic logic circuits in adjacent cycles losing their state due to a premature evaluation is aggravated by the overlapped clock signals. That is, when the next phase clock activates, then the previous phase logic must maintain the necessary precharge state to prevent a false logical evaluation.[39]

In a real chip design using overlapped clocks, the design of the overlap must take into account the inherent clock edge inaccuracy. Thus, the amount of phase time increase is the amount of guaranteed overlap in the clock active states. Equivalently, this

---

37. This is not true of all overlapped systems: the overlap and phase times can and do vary widely by design.

38. See equation (8.68).

39. See, for example, Section 3.2.1 on page 93.

---

means that in such a system the overlap of the clocks must be designed in foreknowledge of the process variation effects on the clock edges.[40]

Furthermore, note that the use of an overlapped system does not preclude the use of time stealing. That is, if a cycle or phase requires additional time beyond the increase, then the same time stealing techniques and rules of Section 8.3.1 remain valid.

8.3.3 Time Stealing for Dynamic Logic in Asymmetric 2-Phase Systems

In an asymmetric two-phase system, time stealing for dynamic logic is identical to the symmetric case.

Directly parallel to the slack borrowing case, asymmetric clocks[41] change the phase times according to the active time split associated with the clock signals. If we apply these times to time stealing, then because the phase delays are equal to the phase times,[42] there is little effect. That is, although the individual phase delays change, the maximum cycle delay remains equal to the sum of the phase delays, or one cycle time in a non-edge adjusted system.

If the extension is made to an asymmetric system to include time stealing, then the maximum amount of time in an n-stage pipeline that can be used by a cycle or phase remains the same as given by equations (8.87) and (8.88). That is, the maximum delay is dependent on the number of stages in the pipeline and the latch delay delays, not the symmetry of the clocking system.[43] Consequently, the rules governing time stealing for dynamic logic in an asymmetric two-phase clocking system are identical to the symmetric case.

8.3.4 Time Stealing in Master-Slave Systems

Time stealing for master-slave logic closely parallels the dynamic logic case.

Master-slave design styles have traditionally been dominant in industry, mostly because of their simple nature.[44] However, because they are by definition edge-triggered systems, they prevent the use of slack borrowing techniques. Consequently, if a pipeline stage in a master-slave system requires additional evaluation time, it must

---

40. See Section 7.2.2.
41. See Figure 8.6 on page 305.
42. See equation (8.65) and (8.66) on page 315.
43. This information can be found on page 320.
44. See Section 5.3.

---

use time stealing techniques to adjust the clock arrival time. A sample falling edge master-slave system is shown in Figure 8.15. In the figure, note that the master-slave system utilizes static logic. Because there is only a single clock, there is no time to hide a precharge phase between machine cycles as in a multi-phase system.[45]

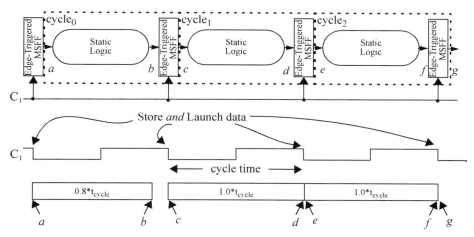

FIGURE 8.15 Sample Master-Slave System using Static Logic

Because there is a only single clock ("C_1") individual phases do not exist. Thus, there are only cycle boundaries to consider. Consequently, there can only be *cycle* time stealing in a master-slave system.

> **Time Stealing Rule:** Because there are no individual phases in a master-slave system, there can only be cycle time stealing.

As in the case for dynamic logic in a two-phase system, dead time penalties are unavoidable. That is, because the system is edge-triggered, a subsequent cycle cannot start its evaluation until a clock arrives. For example, note in Figure 8.15 that point b from cycle$_0$ is available two-tenths of a cycle before the fall of C_1. However, the master-slave flip-flop (MSFF) in cycle$_1$ does not store the data until the next fall of C_1. Thus, any unused time in a cycle is automatically lost.[46]

> **Time Stealing Rule:** Dead time penalties are unavoidable because a master-slave system is edge-triggered.

---

45. See Section 7.5.

46. Unless an early clock signal is available from the clocking system. See Section 8.3.5.

Additionally, the edge-triggered nature of the system means that the minimum and maximum cycle delays of a non-clock adjusted master-slave system are equal to the cycle time, also identical to equations (8.68) and (8.71) for the dynamic case.

> **Time Stealing Rule:** If clock edges are not adjusted, then both the minimum and maximum cycle delay is equal to the cycle time.

Furthermore, to increase a cycle time the clock edge that performs both the storing and launching of the data must be delayed. That is, to gain additional time a cycle must delay the storage of its result by the next master-slave flip-flop. For example, consider what would happen if point d in Figure 8.15 was late. Because the MSFF in cycle$_1$ stores the value on d when C_1 falls, the wrong data will appear on point e unless additional evaluation time for cycle$_1$ is stolen from cycle$_2$. Stealing this time can be easily done by delaying the falling edge of the C_1 clock to the MSFF.

> **Time Stealing Rule:** A cycle delay increase (time stealing) is done by moving (delaying) the store/launch clock edge.

Furthermore, note that cycle$_1$ can only steal time from cycle$_2$, or the next subsequent cycle in the machine pipeline.

> **Time Stealing Rule:** A cycle may steal time only from the next subsequent cycle in the machine pipeline.

As in both the slack borrowing and dynamic logic cases, the sum total of all the delays in an n-stage pipeline machine can be a maximum of n times the cycle time, else the machine will not operate at the desired frequency.[47]

> **Time Stealing Rule:** The sum of all cycle delays across an n-stage pipeline machine can be a maximum of n times the cycle time.

If the clocks are moved in a master-slave system, then, similar to the dynamic case, there is a minimum amount of time a cycle delay must use. That is, if all the latches are moved adjacent to one another, then the machine pipeline will require a minimum amount of time to operate, even though there is no logic present. Each latch will require a minimum time to store and launch data, which is now the cycle delay.

> **Time Stealing Rule:** The minimum cycle delay in a clock adjusted system is equal to the delay of single master-slave flip-flop.

----

47. See equation (8.19) on page 297.

If the minimum delay each MSFF uses is denoted as "t_{MSFF}"[48] then the minimum delay a n-stage pipeline will require is

$$\text{minimum delay across } n \text{ cycles} = n * t_{MSFF}. \tag{8.95}$$

Using equation (8.19) and subtracting equation (8.95), this means that the maximum logic delay in an n-stage master-slave pipeline is

$$\text{maximum } \textit{logic circuity delay} \text{ across } n \text{ cycles} = n * (t_{cycle} - t_{MSFF}). \tag{8.96}$$

Time Stealing Rule: The maximum logic circuitry delay in a clock adjusted n-stage pipeline machine is n times the quantity of the cycle time minus the master-slave flip-flop delay.

If this is used by a single cycle in the pipeline, then the maximum cycle delay is the maximum logic circuitry delay (given by equation (8.96)) plus the delay of the master-slave flip-flop:

$$\text{maximum}\{t_{dcycle}\} = n * (t_{cycle} - t_{MSFF}) + t_{MSFF}. \tag{8.97}$$

Re-writing,

$$\text{maximum}\{t_{dcycle}\} = n * t_{cycle} - (n-1)*(t_{MSFF}). \tag{8.98}$$

Time Stealing Rule: The maximum cycle delay in a clock adjusted n-stage pipeline machine is n times the cycle time minus $n-1$ times the master-slave flip-flop delay.

However, if the flip-flops are connected output-to-input, the designer must be sure that all hold time requirements are met. That is, by moving the flip-flops adjacent to one another, the launch of data from a first automatically means the input to the second has changed. If the new (next) data arrives before the hold time has expired, then it may overwrite the previous value, resulting in a loss of the machine state. Clock edge uncertainty between flip-flops becomes a concern as the clock signal driving each flip-flop may differ. The uncertainty arises from the system-wide clock skew and jitter budget as well as the uncertainty in the clock delaying circuitry.[49]

Time Stealing Rule: In a clock adjusted system, if all the available evaluation time is consumed by a cycle, then the designer must be careful to honor all hold time requirements for the system, adding in clock edge uncertainty from the clock generation and any additional uncertainty introduced by the clock delaying.

---

48. Note that this time includes both the set-up time for the MSFF as well as the clock launch delay time. See Chapter 5 for a more complete discussion on master-slave flip-flops.

49. See Section 7.2.

Additionally, reconsider the effects of point d being late in Figure 8.15 on page 327. If it arrived 10 [ps] later than the falling edge of C_1, then delaying C_1 by that same 10 [ps] would force the MSFF of cycle$_1$ to store the correct information. Thus, if this problem was not discovered during the design time (so as to use stealing techniques) slowing the entire machine clock by 10 [ps] would fix the problem. Note that this is a "once-paid" penalty.[50] This also means that the machine frequency is slowed per equation (8.1).

> **Time Stealing Rule:** In a master-slave system, a system-wide clock signal delay results in only a "once-paid" penalty.

Comparing this back to the slack borrowing discussion, there is again no concept of time averaging for time stealing.[51] That is, any additional time required for an n-stage pipeline cannot be averaged across any number stages.

> **Time Stealing Rule:** Because the system is edge-triggered, time averaging is not possible.

A further note should be made regarding the necessity of time stealing in master-slave systems. From Section 5.2, a master-slave flip-flop consists of two transparent latches connected in series with opposite clocks. If, then, a particular pipeline cycle requires additional time for evaluation and the previous cycle has adequate unused (slack) time, then splitting the master-slave flip-flop into two separate latches with intervening logic will permit the use of slack borrowing instead of time stealing. This can potentially negate the necessity of time stealing.

8.3.5 Time Stealing from a Previous Cycle or Phase

Time stealing from a previous cycle or phase requires an early clock signal and is limited by the earliest time a clock may be obtained from the clocking system.

In some systems, the use of an early clock is possible to start an evaluation cycle or phase earlier in time. The use of such a technique requires the adjustment of clock arrival time(s) at the circuit(s)/latch(es), violating the definition of slack borrowing.

Stealing time from a previous cycle or phase requires that an early clock is available. In many systems, a single clock signal is centrally driven and routed via a balanced tree throughout the chip. At specific intervals the clock is buffered to provide crisp signal edges. At some point, the clock arrives at the circuits and latches to be used, possibly after going through a phase generation block.[52] Consequently, an early clock

---

50. Note that is in contrast to the "twice paid" penalty of the dynamic logic case on page 323.
51. See page 306 for more information on slack borrowing and the effects of time averaging.
52. As would be required in a multi-phase system.

signal can be obtained from the clock tree by tapping into the system at a point up the "branch" from the standard point. In other systems, the clock is routed from a central driver to a chip-wide grid. In this case, there is but a single clock available and no earlier signal can be garnished.[53]

> **Time Stealing Rule:** Time stealing from a previous cycle/phase requires the existence and use of an early clock signal to begin evaluation.

The uniqueness of time stealing from the previous cycle is that the use of an earlier clock signal to begin an evaluation is automatic once the earlier clock signal is garnished. That is, delaying the clock signal is no longer required to steal the necessary additional time. However, the consequence is that the time must come from somewhere, and that time is from the *previous* logic. Thus, if previous time stealing is used, the previous cycle/phase must have slack time to be stolen, and that time must be equal to or greater than the amount needed/stolen.

> **Time Stealing Rule:** Time stealing from a previous cycle/phase can only be performed if the previous cycle has slack time equal to or greater than the amount stolen.

Furthermore, if an earlier clock signal is garnished from the system clock, then its duty cycle will match the system-level clock. In general, the duty cycle of the clocks in the system at the circuits and latches mimic the system-level.[54] If an earlier clock signal is garnished from the central clock tree, then its inactive edge will occur earlier than that of a *normal* clock. Therefore, the active time will not be increased if that clock is used in its raw form. However, if the inactive edge of the early clock is delayed to occur at the same time of a normal clock from the end of tree, then the cycle/phase time of will be increased. Note the similarity to standard time stealing.

> **Time Stealing Rule:** To steal time from a previous cycle/phase, the inactive (falling) edge of the early clock must be delayed as if standard (next) time stealing were performed.

Thus, the amount of time that can be stolen from a previous cycle is limited to the maximum time difference between the early and normal clocks. It is not necessarily a function of the *clocking style*, but a function of the *clock distribution system*.

> **Time Stealing Rule:** The amount of time that can be stolen from a previous cycle is limited to the maximum time difference between the early and normal clocks available from the clock distribution system.

Obviously, the minimum time that can stolen from a previous cycle is zero.

---

53. See Section 7.2.
54. See, for example, Figure 8.1 on page 288.

Time Stealing Rule: The minimum amount of time that can be stolen from a previous cycle is zero.

Additionally, using an early clock may cause one or more outputs to switch *too* early. If the circuit that utilizes the early clock has multiple logic paths, then it possible for an output to switch so early that it violates the hold time of a receiving circuit.

Time Stealing Rule: When stealing time from a previous cycle/phase, special care must be taken to prevent hold time violations on receiving circuits.

In summary, stealing time from a previous cycle or phase is a technique that relies on an early clock to being an early evaluation. This can be done across a cycle boundary, such as two cycles in a master-slave design, or across a phase, such as an L1 to L2 phase boundary. However, the previous cycle or phase relinquishing the time must have sufficient slack to provide. This slack can be a maximum of the amount of time the early clock can be obtained. Furthermore, the clocking system must be compatible wit providing the early clock.

8.3.6 Looping Effects on Time Stealing

Loops (feedback) inside logical pipelines force the application of time stealing rules to each individual loop.

Real chip design environments inevitably require internal looping of paths, regardless of the circuitry and clocking style. That is, a state machine inside a design, such as a microprocessor, nearly always requires input from a previous operation/state to determine the next. This input may come from the next stage in the pipeline or many down the pipeline. Called "loops" because they create return paths in the logical flow, they complicate the analysis of time stealing by creating additional requirements on the system. That is, each individual loop must now be considered a separate machine in itself.

Identical to slack borrowing, each loop in a system introduces an additional constraint on time stealing that can, potentially, limit machine performance. Every loop must comply with the time stealing rules.

Time Stealing Rule: Internal loops must follow the time stealing rules as if they were a machine in themselves.

8.4 Summary

In a chip design, inevitably a logical pipeline partition will require more time than is available, whether a full-cycle or a phase. Depending on the clocking system and the latching structure, obtaining this time is either *slack borrowing* and *time stealing*. Slack borrowing allows a logical partition to utilize time left over (slack time) by the previous partition and requires no clock signal adjustments. Time stealing allows a partition to utilize a portion of the time allotted to the next partition and requires clock signal adjustments. Slack borrowing is ideally suited to static logic in a two-phase system, while time stealing is suited for edge-triggered systems such as dynamic logic in a two-phase system or static logic in a master-slave system. Rules for both borrowing and stealing can be inferred from analyzing the clocking system in conjunction with the particular logic style.

REFERENCES

[8.1]   A. K. Rapp, "A Self-Timing Four Phase Clock Generator," *U.S. Patent #5398001,* May 14, 1995.

[8.2]   R. J. Dupcak, etal, "A Dual-Execution Pipelined Floating-Point CMOS Processor," *Digest of Technical Papers of 1996 IEEE International Solid-State Circuits Conference*, pp. 358-359.

[8.3]   J. P. Uyemura, "Circuit Design for CMOS VLSI," *Kluwer Academic Publishers*, 1992, pp. 275-286.

[8.4]   L. A. Glasser and D. W. Dobberpuhl, "The Design and Analysis of VLSI Circuits, *Addison Wesley Publishing Company Inc., 1985,* pp. 438-443.

[8.5]   M M. Mano, "Computer System Architecture - Second Edition," *Prentice-Hall Inc., New Jersey, 1982,* pp.274-278.

CHAPTER 9

Future Technology

9.1 Introduction

VLSI Technology has traveled a long and challenging road over the last three decades. CMOS process has dominated the last two of those decades due to its low budget for standby power and the success enjoyed by what is largely known as 'scaling theory'.

9.2 Classical Scaling Theory

Classical Scaling Theory, described by Dennard et al. [9.1] has provided the framework about which CMOS technology has advanced. In this scheme, based on a hypothetical prototype technology (e.g.1 μm CMOS) the physical dimensions (width, breadth and depth) of all structures are scaled down in proportion to a factor, κ, $\kappa < 1$, along with all the operating voltages. Together with silicon doping levels which are increased as κ^{-1}, this results in a new, scaled technology which reproduces the function of the prototype, except that the circuit delays are reduced by κ, the power is reduced by κ^2 and the density increased by κ^{-2}. All three of these results, of course, are highly desirable and have driven the painful and expensive progress in lithography forward at a remarkable rate to where today, in 1998, many fabricators are producing 0.25μm-scale technology and are well into the development of 0.18μm.

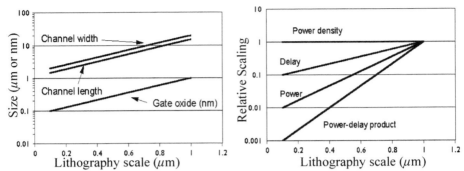

FIGURE 9.1 Scaling of physical dimensions and electrical responses.

This scaling scenario carries with it two theoretical limitations, both described by Dennard et al. as well. First, threshold voltage reduction (scaling) brings with it non-scaling MOSFET sub-threshold conduction, since kT/q is fixed (obviously scaling k, or q is even beyond nobel-prize material, and few customers are willing to move to northern or southern polar regions to scale T). Second, the conductivity of the inter-connects (metal wires) cannot increase as κ^{-1}, and thus the delays associated with wire RC times will not decrease in proportion to the transistor-dominated delays. Both of these impediments will be pursued after a brief review of how process technology scaling has actually proceeded in the industry.

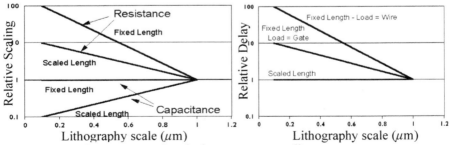

FIGURE 9.2 Interconnect electrical responses to scaling.

9.3 Industry Trends Define Scaling Law

9.3.1 De-Facto Scaling

A de-facto scaling law can be found simply by examining the industry trends over the last fifteen years [9.2], [1.3]. Figure 9.3 shows how power-supply voltage, V_{DD}, has decreased with lithography scaling, and how the key process

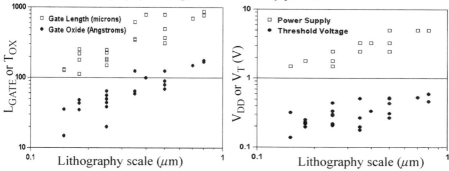

FIGURE 9.3 Published values for gate length, oxide thickness, power supply voltage and threshold voltage of various CMOS technologies.

parameters of gate length and gate oxide (electrical) thickness have been driven. Of particular interest to the circuit and product designer, are the implications of these in device-level responses, as shown in Figure 9.4. The most obvious trend is that

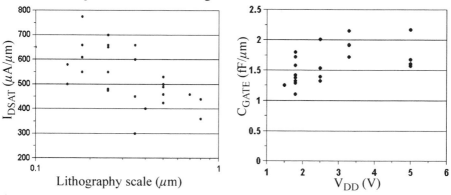

FIGURE 9.4 Published values of n-type MOSFET drive currents and gate capacitances (calculated from gate lengths and oxide thicknesses).

the resultant current drive (normalized to device width) has been slowly creeping upward as the industry has matured, driving higher electric fields across the gate dielectric and across the channel length. Less obvious is the observation that because gate lengths have scaled more-rapidly with V_{DD} reduction than the oxide thickness, the gate capacitance per unit width has been slowly decreasing; this should be contrasted with classical scaling where C_{GATE} is expected to remain constant.

9.4 Challenges Presented by I/S Scaling

9.4.1 V_T Non-scaling

Of greater interest is the actual trend in threshold voltage, shown in Figure 9.5a To interpret this trend one must be reminded of the discussion of Chapter 1 on short-channel effects and threshold voltage. In particular, the V_T at nominal process conditions at 25 $^{\circ}$C will be significantly higher than that found when gate lengths are at the minimum value allowed by process variation and at highest operating temperature. Also drawn in Figure 9.5a are lines indicating that while the process-nominal V_T has been scaling in proportion to V_{DD}, the 'worst-case' low V_T has been nearly constant. This floor value is set by a total chip quiescent current, noise stability, or non-conducting hot-carrier shift limitations, all of which begin to become significant concerns in the vicinity of 100 mV to 200 mV at 25 $^{\circ}$C. This V_T floor will begin to lead to an impasse of sorts for technology scaling, as the extrapolation of nominal V_T reaches that of the worst-case V_T; this is expected to occur at $V_{DD} \sim$ 1.0 - 1.2 V. An empirical fit to the nominal-process V_T trend is shown in Figure 9.5a.

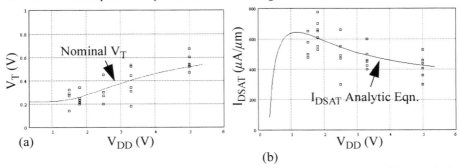

FIGURE 9.5 Threshold Voltage and n-type MOSFET I_{DSAT} trends are fit to analytic current equations. Floor on minimum V_T causes eventual downturn in the fit to drive current.

Using standard analytical equations for drain current one can fit the industry trend well. However, if this trend is continued with the V_T bound suggested, the fit 'predicts' that drive currents will peak at 1 V technology and decrease thereafter (Figure 9.5b). While it would be foolish to suggest that this extrapolation foretells the future of VLSI, it *does* suggest that the industry will undertake some significant changes to avoid such a limitation to future growth.

9.4.2 Gate Lithography

Another hazard on the technology roadmap is evident from the demands on physical gate lengths of the MOSFETs used to drive performance ever higher. The end of conventional optical lithography is signaled by the crossing of images below 400nm, the wavelength of violet light. Mid-Ultraviolet (Mid-UV) lithography is already in manufacturing to support 0.25 μm technologies, using lines from the emission spectrum of ultraviolet light at $\lambda = 248nm$. The *Semiconductor Industry Association Roadmap* [9.4] forecasts gates too small to be resolved even by the next generation (193nm) wavelengths. Many 'tricks' such as off-axis illumination and phase-shift lithography have been proposed to extend the resolution of photo tools beyond their conventional limits. Even with these technologies, eventually, the currently identified photolithographic systems will run out of steam. Contenders for lithography below 100nm include X-ray, Extreme UV and various particle-beam exposure schemes.

9.4.3 Gate Dielectric

The gate oxide thickness data from the industry trend Figure 9.3 currently extends to 3 nm gate oxide for 1.5 V generation technology. It is a straight-forward extrapolation that sub-100 nm CMOS will require thicknesses in the range of 1.5 to 2.5 nm. Analysis by some authors [9.5] forecasts a limit in the 1.5 to 2.0 nm thickness regime due to the onset of direct tunneling from the gate electrode to the MOSFET channel. At such thicknesses this current presents an objectionable drain of standby power. Some work reported [9.6] suggests that prior to the tunnel-current limit, silicon dioxide gate dielectric will be limited by an intrinsic defect generation mechanism, perhaps in before 1 V technology is reached. However much work on alternate gate dielectrics is in progress [9.7] and thus this is not expected to become a scaling issue; new wear-out mechanisms may accompany new dielectrics and so design considerations discussed in Chapter 1 may be altered in the 1 V regime.

9.4.4 Interconnect Delays and Noise

Interconnect delays have never scaled, as pointed out at the beginning of this chapter. This is simply because the conductivity of the interconnect material would have to increase as κ^{-1}. Until the recent [9.8], [9.9] introduction of copper, aluminum had been the mainstay in interconnects. No material with significantly improved conductivity at room temperatures is currently known; thus barring a major breakthrough in new materials (such as high-temperature superconductors), copper interconnects represent a fixed entity that will not scale. The trend in high-performance interconnects has been to selectively scale the vertical dimensions of the interlevel dielectrics and interconnect metal more-slowly than the horizontal dimensions, thus avoiding the severity of interconnect delay degradation that would accompany classical scaling Figure 9.2. This would have resulted in increased interconnect capacitance, however, lower dielectric-constant insulators have mitigated these increases. A similar scenario is expected to play out after the full implementation of copper, beginning with low aspect-ratio interconnects and high dielectric constants with new, lower dielectric-constant insulators being introduced as aspect ratios climb.

9.5 Possible Directions

1.5.1 Faster MOSFETs

Here we explore various emerging technologies that may provide new directions to continue the trend of increasing performance beyond the limits discussed in Section 9.3. While it is not yet clear which, if any, of these schemes may provide the thrust for this next trend of advances, it is clear that the demand for such a continuation assures discovery, if at all possible.

Multiple Threshold Voltages: Selective use of low V_T devices in critical paths has been introduced in both high performance [9.10] and low power [9.11] applications. A budget of subthreshold current is spent on those devices that must be low-V_T to provide small circuit delays, while less-drive-critical circuits (such as SRAM caches) use higher V_T to keep standby current under control. This technique is likely to become more prevalent as pressure from voltage scaling makes lower V_T necessary.

<u>Silicon-Germanium MOSFETs</u>: Mobility improvements, despite the limiting effects of saturation velocity in high-performance MOSFETs, still can return significant improvements in switching speed [9.12] Figure 9.6 shows the potential benefits. One technique being explored exploits the fact that the mobility of holes in SiGe crystals is considerably higher than that in pure silicon [9.13]. Figure 9.7 shows a simplified

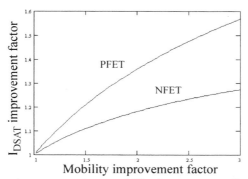

FIGURE 9.6 I_{DSAT} improvement vs. mobility improvement for.25 μm MOSFETs.

version of the structures used to integrate SiGe P-type MOSFETs into CMOS technology. Buffer layers of lower-fraction Ge-alloyed silicon relieve stress to allow a gradual transition to an alloy with ~50% Ge in the active channel region. Since high-quality gate oxide cannot be grown in SiGe, a layer of pure silicon tops off the channel to be used to form the gate oxide. Inversion-layer holes will remain in the SiGe because of the higher valence-band energy in SiGe over Si (provided the gate bias does not become too extreme).

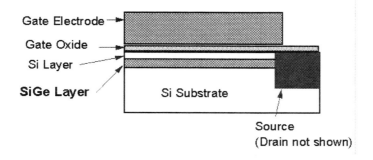

FIGURE 9.7 Schematic cross-section of SiGe heterojunction p-type MOSFET integrated into silicon technology.

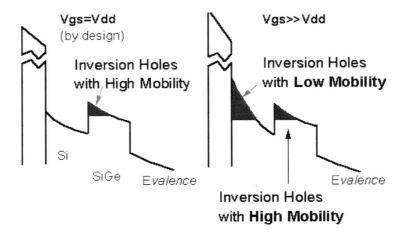

FIGURE 9.8 Band diagram showing operation of SiGe p-type MOSFET shown in Figure 1.7.

A scheme to increase electron mobility, also employing SiGe, has been demonstrated on silicon [1.14]. In this case, SiGe is grown on a Si substrate and then more pure Si is grown on top of that layer to form a strained layer of pure silicon at the top. Under the correct conditions, the mobility of the electrons can be increased considerably.

Silicon on Insulator: Isolating every MOSFET individually in oxide is the goal of this technology. It offers near-zero junction capacitance and the potential to eliminate body effects from circuit delays. Challenges in manufacture of high-quality silicon-on-insulator wafers and in design of circuits and technologies that can manage the affects from floating bodies in MOSFETs need be overcome for this technology to enter the mainstream.

Dynamic Threshold CMOS: One novel suggestion [9.18] to circumvent finite V_T requirements in CMOS is to make the V_T *dynamic*. By connecting the bodies of MOSFETs to their own gates the threshold voltage itself becomes a function of the gate drive. By properly designing the body doping profile the V_T can be made, for example, zero when $V_{GATE} = V_{DD}$ for high overdrive, while maintaining a conventional value (perhaps 0.25 V) when $V_{GATE} = 0V$, to keep off-currents low. Of course the power supply voltage is limited to less than the forward-conduction voltage of p-n diodes in order to avoid excessive currents from body to source; hence this technology can only

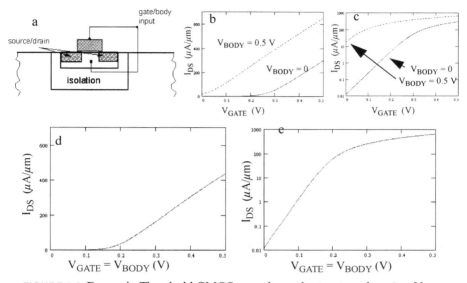

FIGURE 9.9 Dynamic Threshold CMOS, a. schematic structure, b, c, I_{DS}-V_{GATE}.

come into consideration for $V_{DD}<.6V$. It has also been pointed out [9.16] that capacitance associated with the body to the source and drain must be kept minimal and thus it remains to be demonstrated that such a structure with good performance is really feasible.

1.5.2 Challenges of the Unknown

Viewing the future, other, more-global issues appear on the horizon. We mention a few of these to give some flavor of the unknowns than may lie ahead.

Variable Hold Time / Variable Retention Time Defects: In DRAM technology, workers have reported a form of leakage which appears to "switch" on and off as a function of time in a random manner [9.17]. As scaling pushes logic to smaller scales, one may reach the stage where such a mechanism may be capable of intermittently disturbing logic states. Testing such a phenomenon out would seem beyond reasonable possibility and new approaches would need be found.

Atomic-Scale Effects: At gate lengths in the vicinity of 20-50 nm, silicon devices will become noticeable influenced by the atomic nature of matter [9.18]. Number fluctuations of dopant atoms, finite lengths of wave-functions of inversion layers, and other non-scaling atomic properties appear to be 'dead end' signs for the CMOS scaling juggernaut. It will be interesting to see if this limit is effectively a real one.

Cost: As density and performance geometrically improve, so due the costs of tools and fabricators. This continually brings up the speculation that few will be able to afford the enormous outlays of capital to enter VLSI fabrication as this spiral continues. Thus far such predictions have failed to materialize. Like a "bull" stock market, however, some triggering event may finally cause a radical change in the value and investment climate surrounding VLSI manufacturing. The authors sincerely wish that such is not the case.

REFERENCES

[9.1]   R.H.Dennard, F.H.Gaensslen, H.Yu, V.L.Rideout, E.Bassous and A.R. LeBlanc, "Design of ion-implanted MOSFETs with very small dimensions," IEEE *J. Solid State Circuits*, SC-9, 25 (1974)1

[9.2]   E.J.Nowak, "Ultimate CMOS ULSI Performance," IEEE *Technical Digest of the IEDM* pp.115-118 (1993)

[9.3]   Y.Taur and E.J.Nowak, "CMOS Devices below 0.1μm: How High Will Performance Go?", IEEE *Technical Digest of the IEDM* pp.215-218 (1997)

[9.4]   "The National Technology Roadmap for Semiconductors Technology Needs," 1997 Edition, Semiconductor Industries Association p.27 (1997)

[9.5]   S.-H.Lo, D.A.Buchanan, Y.Taur and W.Wang, "Quantum-Mechanical Modeling of Electron Tunneling Current from the Inversion Layer of Ultra-Thin-Oxide nMOSFET's," IEEE *Electron Device Letters* Vol.18, No.5, pp. 209-211 (1997)

[9.6]   R.Degraeve, G. Groeseneken, R. Bellens, M. Depas, and H.E. Maes, "A consistent model for the thickness dependence of intrinsic breakdown in ultra-tin oxides," IEEE *Technical Digest of the IEDM* pp863-866 (1995)

[9.7]   T.P.Ma, "Making Silicon Nitride a Viable Gate Dielectric," IEEE *Transactions on Electron Devices* 45, No.3, pp.680-690 (1998)

[9.8] D.Edelstein, J.Heidenreich, R.Goldblatt, W.Cote, C.Uzoh, N.Lustig, P.Roper, T.McDevitt, W.Motsiff, A.Simon, J.Dukovic, R.Wachnik, H.Rathore, R.Schulz, L.Sue, S.Luce, and J.Slattery, "Full Copper Wiring in a Sub-0.25 μm CMOS ULSI Technology," *IEEE Technical Digest of the IEDM*, pp.773-776, (1997)

[9.9] S.Venkatesan et al., "A High Performance 1.8V, 0.20 μm CMOS Technology with Copper Metallization," IEEE *Technical Digest of the IEDM* pp. 769-772, (1997)

[9.10] N.Rohrer et al. "A 480Mhz RISC Microprocessor in a 0.12 μm-L_{EFF} CMOS Technology with Copper Interconnects," *Proceedings of the 1998 International Solid State Circuits Conference*, pp.240-241 (1998)

[9.11] W.Lee et al., "A 1V DSP for Wireles Communications," *Proceedings of the 1998 International Solid State Circuits Conference*, pp.92-93 (1997)

[9.12] Y.Taur and E.J.Nowak, op cit

[9.13] D.K.Nayak, J.C.S.Woo, G.K.Yabiku, J.S.Park, K.L.Wang, and K.P.MacWilliams, "Enhancement-Mode Quantum-Well Ge_XSi_{1-X} PMOS," IEEE *Electron Device Letters*, Vol.12, No.4, pp.154-156 (1991)

[9.14] H. Jorke and H. J. Herzoz, "Mobility enhancement in modulation doped Si-$Si_{1-X}Ge_X$ superlattice grown by molecular beam epitaxy," Proc. 1st Int. Symp. on Silicon Molecular Beam Epitaxy, Electro.-Chem. Soc., pp.325-335 (1985)

[9.15] F.Assaderaghi, D. Sinitsky, S. Parke, J. Boker, P. Ko and C. Hu, "A Dynamic Threshold Voltage MOSFET (DTCMOS) for Ultra-Low Voltage Operation," IEEE *Technical Digest of the IEDM*, pp.809-812 (1994)

[9.16] C.Wann, F.Assaderaghi, R.Dennard, C.Hu, G.Shahidi and Y.Taur, "Channel Profile Optimization and Device Design for Low-Power High-Performance Dynamic-Threshold MOSFET," IEEE *Technical Digest of the IEDM*, pp.113-116 (1996)

[9.17] D.S.Yaney, C.Y.Lu, R.A.Kohler, M.J.Kelly and J.T.Nelson, "A Meta-Stable Leakage Phenomenon in DRAM Charge Storage - Variable Hold Time," IEEE *Technical Digest of the IEDM*, pp.336-339, (1987)

[9.18]   H.Iwai and H.S.Momose, "Technology towards low power / low voltage and scaling of MOSFETs," *Microelectronic Engineering* 39 (1997) pp.7-30 (1997)

INDEX

snapback 219
soft error rates 31
Soft Errors 168
spacer 19
SRCMOS 200
stacked-device driver 231
Static CMOS 54, 134, 160
Static Combinatorial CMOS Logic 55
Static evaluation 97
static latch 195
static logic 287, 294
Substrate hot carrier 17
switch point 137
Switched Output Differential Structure (SODS) 123
symmetric 288, 306, 313
System Clock 289

T

TEOS 36
Threshold variation 13
time averaging 309, 323, 330
Time stealing 285, 286, 312
titanium 23
transmission gates 145, 165
Transmission Line 167, 212
Transparent 289, 294, 301
transparent latches 287
tree 59
True Single Phase Clocking 184, 186, 203
two-phase 286, 305, 306, 312, 313

U

undershoot 97
unity gain point 137

V

Voltage Translation 230, 240

Z

Zipper Domino 110